Mastering
Python Design Patterns

Craft essential Python patterns by following core
design principles

Kamon Ayeva

Sakis Kasampalis

Mastering Python Design Patterns

Copyright © 2024 Packt Publishing

Group Product Manager: Kunal Sawant
Publishing Product Manager: Samriddhi Murarka
Book Project Manager: Manisha Singh
Senior Editor: Nithya Sadanandan
Technical Editor: Vidhisha Patidar
Copy Editor: Safis Editing
Proofreader: Nithya Sadanandan
Indexer: Tejal Soni
Production Designer: Nilesh Mohite
DevRel Marketing Coordinator: Shrinidhi Manoharan

First published: January 2015
Second edition: August 2018
Third edition: May 2024

Production reference: 1230524

Published by Packt Publishing Ltd.
Grosvenor House
11 St Paul's Square
Birmingham
B3 1RB, UK

ISBN 978-1-83763-961-8

www.packtpub.com

I would like to thank my parents for their love and support.

– Kamon Ayeva

Contributors

About the authors

Kamon Ayeva is a seasoned Python expert with over two decades of experience in the technology sector. As the founder of Content Gardening Studio, a consulting and custom development services firm, he specializes in web development, data, and AI, delivering top-notch Python solutions to clients globally. A trusted educator, Kamon has trained numerous developers, solidifying his reputation as an authority in the Python community. He is also the co-author of the previous edition of Mastering Python Design Patterns. On social media, you can find him on Twitter under the handle `@kamon`, where he continues to share invaluable insights and trends in Python and software design.

Sakis Kasampalis is a software architect living in the Netherlands. He is not dogmatic about particular programming languages and tools; his principle is that the right tool should be used for the right job. One of his favorite tools is Python because he finds it very productive. Sakis was also the technical reviewer of Mastering Object-oriented Python and Learning Python Design Patterns, published by Packt Publishing.

About the reviewers

Fréjus L. O. Adjé is a software engineer with more than 10 years of industry experience. Holding a degree in Computer Network and Internet Engineering, he is highly regarded for his proficiency in creating dynamic web applications. With nearly 7 years of focused expertise in Python, he has successfully led teams and delivered innovative solutions. Committed to lifelong learning, he actively seeks to stay ahead in the rapidly evolving technology landscape while sharing his expertise by mentoring new developers. Off-duty, Fréjus enjoys chess, design, and football, dedicating his free time to chess matches, exploring trends, and playing or watching football, which broadens his global perspective and enriches his software development approach.

Encolpe DEGOUT has been a developer and an open-source advocate since 1998. He started using Python and Zope in 2002 and then became involved in Nuxeo CPS and Plone CMS. He promotes the open-source ecosystem by organizing conferences and giving courses. Besides these involvements, he participates in automating data integration in a search engine specialized in European media and worldwide press.

Gianguglielmo Calvi is a computer scientist and knowledge manager, founder of Heuristica, and co-founder of EnQu Ideation. With a strong background in programming in C/C++ and Python, he has over two decades of experience in international projects. He currently serves as a Senior Knowledge Management Systems Expert consultant at the Green Growth Knowledge Partnership. His career includes roles as a researcher in Cognitive Science and AI at ISTC-CNR and as a knowledge manager at the International Labour Organization, UN/CEFACT, UNDSS, WHO EUROPE, and Voolinks. Gianguglielmo holds a master's degree in computer science from the University of Pisa, a Knowledge Management certification from the IKF Institute, and various international certifications.

Table of Contents

Part 1: Start with Principles

1

Foundational Design Principles   3

2

SOLID Principles   23

Part 2: From the Gang of Four

3

Creational Design Patterns   43

4

Structural Design Patterns   77

5

Behavioral Design Patterns   117

Part 3: Beyond the Gang of Four

6

Architectural Design Patterns   163

7

Concurrency and Asynchronous Patterns   187

8

Performance Patterns   205

9

Distributed Systems Patterns   221

Preface

Explore the world of design principles and design patterns in the context of the Python programming language with this comprehensive guide. Learn about classic and modern design patterns and how to use them to solve problems you encounter daily as a Python developer or software architect.

With code examples, real-world case studies, and detailed solution implementations, this book is a must-read for Python developers looking to elevate their coding skills. Co-authored by a Python expert with over two decades of experience, this new edition expands the scope to cover more design pattern categories. Gain insights into creational, structural, behavioral, architectural, and other important patterns for modern software design, such as concurrency, asynchronous, and performance patterns. Learn how to apply these patterns in various domains like event handling systems, concurrency, distributed systems, and testing. The book also presents Python anti-patterns, helping you avoid common pitfalls.

Whether you're developing user interfaces, web applications, APIs, data pipelines, or AI models, this book equips you with the knowledge to build robust and maintainable software.

This book adopts a hands-on approach, providing code examples for each design pattern. Each chapter includes step-by-step instructions to test the code, making it an interactive learning experience. Where applicable, for each design principle or pattern, the book presents at least one real-world example, which may or may not be Python-based, and at least one Python-based example.

Who this book is for

This book is for Python developers looking to deepen their understanding of design patterns and how they can be applied to various types of projects. With a focus on intermediate and advanced Python programmers, the book also includes introductory chapters that make it accessible for those who are relatively new to the language. Whether you're a web developer, data engineer, or AI specialist, this book offers valuable insights into the best practices for software design, backed by real-world examples and decades of experience. It's also an excellent resource for software architects and team leaders who want to improve code quality and maintainability across their projects.

What this book covers

Chapter 1, Foundational Design Principles, covers principles of encapsulation, composition, programming to interfaces, and loose coupling to help you create more adaptable and maintainable systems.

Chapter 2, SOLID Principles, Provides guidelines for designing robust, maintainable, and scalable software. Each of these principles contributes to creating clean and adaptable code.

Chapter 3, Creational Design Patterns, explores patterns that help manage object creation by controlling which classes to instantiate.

Chapter 4, Structural Design Patterns, provides insights into patterns that facilitate the design process by identifying simple ways to establish relationships between entities. This chapter delves into six essential structural patterns, providing you with the skills to structure your code efficiently and elegantly.

Chapter 5, Behavioral Design Patterns, shares patterns that focus on the interactions and responsibilities of objects, promoting effective communication and flexible assignment of responsibilities. This chapter explores key patterns such as Strategy, Observer, and Command, demonstrating how they streamline object collaboration and enhance the adaptability of code.

Chapter 6, Architectural Design Patterns, delves into patterns that provide templates for solving common architectural problems, facilitating the development of scalable, maintainable, and reusable systems.

Chapter 7, Concurrency and Asynchronous Patterns, explores patterns that help you develop applications that are both fast and user-friendly, particularly in environments with heavy I/O operations or significant computational work.

Chapter 8, Performance Patterns, provides guidance on patterns that address common bottlenecks and optimization challenges, offering proven methodologies to improve execution time, reduce memory usage, and scale effectively.

Chapter 9, Distributed Systems Patterns, shows patterns that empower developers to architect robust distributed systems, from managing communication between nodes to ensuring fault tolerance and consistency.

Chapter 10, Patterns for Testing, presents patterns that help in isolating components, making tests more reliable, and promoting code reusability.

Chapter 11, Python Anti-Patterns, explores common programming practices that, while not necessarily wrong, often lead to less efficient, less readable, and/or less maintainable code. You will learn to understand and avoid these pitfalls.

To get the most out of this book

Use a machine with a recent version of Windows, Linux, or macOS.

Install Python 3.12. It is also useful to create a virtual environment from your Python installation so that when you add third-party modules required for following some of the chapters, you do not end up polluting your global Python. This is a fundamental best practice for productivity with Python, and you will find many resources on the Internet that explain how to do this.

Install and use Docker on your machine. This will help with the requirement of some external software services or tools, such as LocalStack (used in *Chapter 6*) and the Redis server (used in *Chapter 8*).

Software/hardware covered in the book	Operating system requirements
Python 3.12	Windows, macOS, or Linux
MyPy 1.10.0	
Docker	
Redis-server 6.2.6	
LocalStack 3.4.0	

If you are using the digital version of this book, we advise you to type the code yourself or access the code from the book's GitHub repository (a link is available in the next section). Doing so will help you avoid any potential errors related to the copying and pasting of code.

Download the example code files

You can download the example code files for this book from GitHub at https://github.com/ PacktPublishing/Mastering-Python-Design-Patterns-Third-Edition. If there's an update to the code, it will be updated in the GitHub repository.

We also have other code bundles from our rich catalog of books and videos available at https:// github.com/PacktPublishing/. Check them out!

Conventions used

There are a number of text conventions used and formatting specificities throughout this book.

Most of the code has been automatically formatted

The formatting has been done using the Black tool, as is commonly done by Python developers for productivity reasons. So it might not look exactly like the code you would write yourself. But it is valid; it is a PEP 8-compliant code. The goal is to improve the readability of the code snippets.

So, some code snippets in the code files as well as in the book's pages may look like the following:

```
State = Enum(
    "State",
    "NEW RUNNING SLEEPING RESTART ZOMBIE",
)
```

Another example might be the following:

```
msg = (
    f"trying to create process '{name}' "
    f"for user '{user}'"
```

```
  )
  print(msg)
```

The code snippets in the book's pages may be shortened

To improve readability, when there is a documentation string (docstring) for a function or class, and it is too long, we remove it from the code snippet in the book.

When some code (class or function) is too long to display on the chapter's pages, we may shorten it, and refer the reader to the complete code in the file.

> **Note**
> In case of an issue with long commands, which are spread across several lines (with the '/' character as separator), you can reformat the long command text, removing the '/' character, to make sure that the command is correctly interpreted in the terminal.

Other conventions

`Code in text`: Indicates code words in text, database table names, folder names, filenames, file extensions, pathnames, dummy URLs, user input, and Twitter handles. Here is an example: "Define the `Logger` interface with a `log` method."

A block of code is set as follows:

```
class MyInterface(ABC):
    @abstractmethod
    def do_something(self, param: str):
        pass
```

Any command-line input or output is written as follows:

```
python3.12 -m pip install --user mypy
```

Bold: Indicates a new term, an important word, or words that you see onscreen. For instance, words in menus or dialog boxes appear in bold. Here is an example: "It is one of the core concepts in **object-oriented programming OOP** that enables a single interface to represent different types."

> **Tips or important notes**
> Appear like this.

Get in touch

Feedback from our readers is always welcome.

General feedback: If you have questions about any aspect of this book, email us at customercare@packtpub.com and mention the book title in the subject of your message.

Errata: Although we have taken every care to ensure the accuracy of our content, mistakes do happen. If you have found a mistake in this book, we would be grateful if you would report this to us. Please visit www.packtpub.com/support/errata and fill in the form.

Piracy: If you come across any illegal copies of our works in any form on the internet, we would be grateful if you would provide us with the location address or website name. Please contact us at copyright@packt.com with a link to the material.

If you are interested in becoming an author: If there is a topic that you have expertise in and you are interested in either writing or contributing to a book, please visit authors.packtpub.com.

Share your thoughts

Once you've read *Mastering Python Design Patterns*, we'd love to hear your thoughts! Scan the QR code below to go straight to the Amazon review page for this book and share your feedback.

https://packt.link/r/1837639612

Your review is important to us and the tech community and will help us make sure we're delivering excellent quality content.

Download a free PDF copy of this book

Thanks for purchasing this book!

Do you like to read on the go but are unable to carry your print books everywhere?

Is your eBook purchase not compatible with the device of your choice?

Don't worry, now with every Packt book you get a DRM-free PDF version of that book at no cost.

Read anywhere, any place, on any device. Search, copy, and paste code from your favorite technical books directly into your application.

The perks don't stop there, you can get exclusive access to discounts, newsletters, and great free content in your inbox daily

Follow these simple steps to get the benefits:

1. Scan the QR code or visit the link below

https://packt.link/free-ebook/9781837639618

2. Submit your proof of purchase
3. That's it! We'll send your free PDF and other benefits to your email directly

Part 1:
Start with Principles

This first part introduces you to the foundational software design principles and the S.O.L.I.D. principles that build upon them. This part includes the following chapters:

- *Chapter 1, Foundational Design Principles*
- *Chapter 2, SOLID Principles*

1
Foundational Design Principles

Design principles form the foundation of any well-architected software. They serve as the guiding light that helps developers navigate the path of creating maintainable, scalable, and robust applications while avoiding the pitfalls of bad design.

In this chapter, we will explore the core design principles that all developers should know and apply in their projects. We will explore four foundational principles. The first one, *Encapsulate What Varies*, teaches you how to isolate the parts of your code that are subject to change, making it easier to modify and extend your applications. Next, *Favor Composition*, makes you understand why it's often better to assemble complex objects from simple ones rather than inheriting functionalities. The third one, *Program to Interfaces*, shows the power of coding to an interface rather than to a concrete class, enhancing flexibility and maintainability. Finally, with the *Loose Coupling* principle, you will grasp the importance of reducing dependencies between components, making your code easier to refactor and test.

In this chapter, we're going to cover the following main topics:

- Following the "Encapsulate What Varies" principle
- Following the "Favor Composition Over Inheritance" principle
- Following the "Program to Interfaces, Not Implementations" principle
- Following the "Loose Coupling" principle

By the end of this chapter, you'll have a solid understanding of these principles and how to implement them in Python, setting the foundation for the rest of the book.

Technical requirements

For the chapters in this book, you will need a running Python 3.12 environment, or for some exceptional cases in some chapters, 3.11.

In addition, install the Mypy static type checker (`https://www.mypy-lang.org`) by running the following:

```
python3.12 -m pip install --user mypy
```

The examples are available in the GitHub repository here: `https://github.com/PacktPublishing/Mastering-Python-Design-Patterns-Third-Edition`

> **About the Python executable**
>
> Throughout the book, we will reference the Python executable for executing code examples as `python3.12` or `python`. Adapt that to your specific environment, practice and/or workflow.

Following the Encapsulate What Varies principle

One of the most common challenges in software development is dealing with change. Requirements evolve, technologies advance, and user needs also change. Therefore, it is crucial to write code that can adapt without causing a ripple effect of modifications throughout your program or application. This is where the principle of *Encapsulate What Varies* comes into play.

What does it mean?

The idea behind this principle is straightforward: isolate the parts of your code that are most likely to change and encapsulate them. By doing so, you create a protective barrier that shields the rest of your code from these elements that are subject to change. This encapsulation allows you to make changes to one part of your system without affecting others.

Benefits

Encapsulating what varies provides several benefits, mainly the following:

- **Ease of maintenance**: When changes are needed, you must only modify the encapsulated parts, reducing the risk of introducing bugs elsewhere
- **Enhanced flexibility**: Encapsulated components can be easily swapped or extended, providing a more adaptable architecture
- **Improved readability**: By isolating varying elements, your code becomes more organized and easier to understand

Techniques for achieving encapsulation

As we introduced, encapsulation helps in data hiding and exposing only the necessary functionalities. Here, we will present key techniques that enhance encapsulation in Python: polymorphism and the *getters* and *setters* techniques.

Polymorphism

In programming, polymorphism allows objects of different classes to be treated as objects of a common superclass. It is one of the core concepts in **object-oriented programming OOP** that enables a single interface to represent different types. Polymorphism allows for implementing elegant software design patterns, such as the strategy pattern, and it's a way to implement clean, maintainable code in Python.

Getters and Setters

These are special methods in a class that enable controlled access to attribute values. *getters* allow reading the values of attributes and *setters* enable modifying them. By using these methods, you can add validation logic or side effects such as logging, thus adhering to the principles of encapsulation. They provide a way to control and protect the state of an object and are particularly useful when you want to encapsulate complex attributes that are derived from other instance variables.

And there is more. To complement the *getters* and *setters* technique, Python offers a more elegant approach known as the *property* technique. This is a built-in feature of Python that allows you to convert attribute access into method calls seamlessly. With properties, you can ensure that an object retains its internal state against incorrect or harmful manipulation without having to explicitly define *getter* and *setter* methods.

The `@property` decorator allows you to define a method that is automatically invoked when an attribute is accessed, effectively serving as a *getter*. Similarly, the `@attribute_name.setter` decorator allows you to define a method that acts as a *setter*, invoked when you attempt to change the value of an attribute. This way, you can embed validation or other actions directly within these methods, making the code more clean.

By using the *property* technique, you can achieve the same level of data encapsulation and validation as with traditional *Getters* and *Setters* but in a way that is more aligned with Python's design philosophy. It allows you to write code that is not just functional but also clean and easy to read, enhancing both encapsulation and the overall quality of your Python programs.

Next, we will better understand these techniques through examples.

An example – encapsulating using polymorphism

Polymorphism is a powerful way to achieve encapsulation of varying behavior. Let's see that with an example of a payment processing system where the payment method option can vary. In such a case, you might encapsulate each method of payment in its own class:

1. You would first define the base class for payment methods, providing a `process_payment()` method that each specific payment method will implement. This is where we encapsulate what varies—the payment processing logic. That part of the code will be as follows:

```python
class PaymentBase:
    def __init__(self, amount: int):
        self.amount: int = amount

    def process_payment(self):
        pass
```

2. Next, we introduce the `CreditCard` and `PayPal` classes, inheriting from `PaymentBase`, each providing their own implementation of `process_payment`. This is a classic way of polymorphism, as you can treat `CreditCard` and `PayPal` objects as instances of their common superclass. The code is as follows:

```python
class CreditCard(PaymentBase):
    def process_payment(self):
        msg = f"Credit card payment: {self.amount}"
        print(msg)


class PayPal(PaymentBase):
    def process_payment(self):
        msg = f"PayPal payment: {self.amount}"
        print(msg)
```

3. To make it possible to test the classes we just created, let's add some code, calling `process_payment()` for each object. The beauty of polymorphism is evident when you use these classes, as follows:

```python
if __name__ == "__main__":
    payments = [CreditCard(100), PayPal(200)]
    for payment in payments:
        payment.process_payment()
```

The complete code for this example (`ch01/encapsulate.py`) is as follows:

```python
class PaymentBase:
    def __init__(self, amount: int):
```

```
        self.amount: int = amount

    def process_payment(self):
        pass


class CreditCard(PaymentBase):
    def process_payment(self):
        msg = f"Credit card payment: {self.amount}"
        print(msg)


class PayPal(PaymentBase):
    def process_payment(self):
        msg = f"PayPal payment: {self.amount}"
        print(msg)


if __name__ == "__main__":
    payments = [CreditCard(100), PayPal(200)]
    for payment in payments:
        payment.process_payment()
```

To test the code, run the following command:

```
python3.12 ch01/encapsulate.py
```

You should get the following output:

```
Credit card payment: 100
PayPal payment: 200
```

As you can see, when the payment method changes, the program adapts to produce the expected outcome.

By encapsulating what varies—here, the *payment method*—you can easily add new options or modify existing ones without affecting the core payment processing logic.

An example – encapsulating using a property

Let's define a Circle class and show how to use Python's @property technique to create a *getter* and a *setter* for its radius attribute.

Note that the underlying attribute would actually be called _radius, but it is hidden/protected behind the *property* called radius.

Let's write the code step by step:

1. We start by defining the `Circle` class with its initialization method, where we initialize the `_radius` attribute as follows:

```
class Circle:
    def __init__(self, radius: int):
        self._radius: int = radius
```

2. We add the radius property: a `radius()` method where we return the value from the underlying attribute, decorated using the `@property` decorator, as follows:

```
@property
def radius(self):
    return self._radius
```

3. We add the radius setter part: another `radius()` method where we do the actual job of modifying the underlying attribute, after a validation check, since we do not want to allow a negative value for the radius; this method is decorated by the special `@radius.setter` decorator. This part of the code is as follows:

```
@radius.setter
def radius(self, value: int):
    if value < 0:
        raise ValueError("Radius cannot be negative!")
    self._radius = value
```

4. Finally, we add some lines that will help us test the class, as follows:

```
if __name__ == "__main__":
    circle = Circle(10)
    print(f"Initial radius: {circle.radius}")

    circle.radius = 15
    print(f"New radius: {circle.radius}")
```

The complete code for this example (`ch01/encapsulate_bis.py`) is as follows:

```
class Circle:
    def __init__(self, radius: int):
        self._radius: int = radius

    @property
    def radius(self):
        return self._radius
```

```
    @radius.setter
    def radius(self, value: int):
        if value < 0:
            raise ValueError("Radius cannot be negative!")
        self._radius = value


if __name__ == "__main__":
    circle = Circle(10)
    print(f"Initial radius: {circle.radius}")

    circle.radius = 15
    print(f"New radius: {circle.radius}")
```

To test the example, run the following command:

```
python3.12 ch01/encapsulate_bis.py
```

You should get the following output:

```
Initial radius: 10
New radius: 15
```

In this second example, we saw how we can encapsulate the circle's radius component so that we can change the technical aspects if needed, without breaking the class. For example, the validation code for the *setter* can evolve. We can even change the underlying attribute, _radius, and the behavior for the user of our code will remain unchanged.

Following the Favor Composition Over Inheritance principle

In OOP, it's tempting to create complex hierarchies of classes through inheritance. While inheritance has its merits, it can lead to tightly coupled code that is hard to maintain and extend. This is where the principle of *Favor Composition Over Inheritance* comes into the picture.

What does it mean?

This principle advises that you should prefer composing objects from simpler parts to inheriting functionalities from a base class. In other words, build complex objects by combining simpler ones.

Benefits

Choosing composition over inheritance offers several advantages:

- **Flexibility**: Composition allows you to change objects' behavior at runtime, making your code more adaptable

- **Reusability**: Smaller, simpler objects can be reused across different parts of your application, promoting code reusability

- **Ease of maintenance**: With composition, you can easily swap out or update individual components without affecting the overall system, avoiding border effects

Techniques for composition

In Python, composition is often achieved through OOP by including instances of other classes within a class. This is sometimes referred to as a "has-a" relationship between the class that is being composed and the classes that are being included. Python makes it particularly easy to use composition by not requiring explicit type declarations. You can include other objects by simply instantiating them in the class's __init__ method or by passing them as parameters.

An example – compose a car using the engine

In Python, you can use composition by including instances of other classes within your class. For example, consider a Car class that includes an instance of an Engine class:

1. Let's first define the Engine class as follows, with its start method:

```
class Engine:
    def start(self):
        print("Engine started")
```

2. Then, let's define the Car class as follows:

```
class Car:
    def __init__(self):
        self.engine = Engine()

    def start(self):
        self.engine.start()
        print("Car started")
```

3. Finally, add the following lines of code to create an instance of the Car class and call the start method on that instance, when this program is executed:

```
if __name__ == "__main__":
    my_car = Car()
    my_car.start()
```

The complete code for this example (ch01/composition.py) is as follows:

```
class Engine:
    def start(self):
        print("Engine started")


class Car:
    def __init__(self):
        self.engine = Engine()

    def start(self):
        self.engine.start()
        print("Car started")


if __name__ == "__main__":
    my_car = Car()
    my_car.start()
```

To test the code, run the following command:

```
python3.12 ch01/composition.py
```

You should get the following output:

```
Engine started
Car started
```

As you can see in this example, the Car class is composed of an Engine object, thanks to the self.engine = Engine() line, allowing you to easily swap out the engine for another type without altering the Car class itself.

Following the Program to Interfaces, Not Implementations principle

In software design, it's easy to get caught up in the specifics of how a feature is implemented. However, focusing too much on implementation details can lead to code that is tightly coupled and difficult to modify. The principle of *Program to Interfaces, Not Implementations* offers a solution to this problem.

What does it mean?

An interface defines a **contract** for classes, specifying a set of methods that must be implemented.

This principle encourages you to code against an interface rather than a concrete class. By doing so, you untie your code from the specific classes that provide the required behavior, making it easier to swap or extend implementations without affecting the rest of the system.

Benefits

Programming to interfaces offers several benefits:

- **Flexibility**: You can easily switch between different implementations without altering the code that uses them

- **Maintainability**: Losing your code from specific implementations makes it easier to update or replace components

- **Testability**: Interfaces make it simpler to write unit tests, as you can easily mock the interface during testing

Techniques for interfaces

In Python, *interfaces* can be implemented using two primary techniques: **abstract base classes (ABCs)** and protocols.

Abstract base classes

ABCs, provided by the abc module, allow you to define *abstract methods* that must be implemented by any concrete (i.e., non-abstract) subclass.

Let's understand this concept with an example, where we will define an abstract class (for an interface) and then use it:

1. First, we need to import the ABC class and the abstractmethod decorator function as follows:

   ```
   from abc import ABC, abstractmethod
   ```

2. Then, we define the interface class as follows:

```
class MyInterface(ABC):
    @abstractmethod
    def do_something(self, param: str):
        pass
```

3. Now, define a concrete class for that interface; it inherits from the interface class and provides an implementation for the do_something method as follows:

```
class MyClass(MyInterface):
    def do_something(self, param: str):
        print(f"Doing something with: '{param}'")
```

4. Add the following lines for testing purposes:

```
if __name__ == "__main__":
    MyClass().do_something("some param")
```

The complete code (ch01/abstractclass.py) is as follows:

```
from abc import ABC, abstractmethod


class MyInterface(ABC):
    @abstractmethod
    def do_something(self, param: str):
        pass


class MyClass(MyInterface):
    def do_something(self, param: str):
        print(f"Doing something with: '{param}'")


if __name__ == "__main__":
    MyClass().do_something("some param")
```

To test the code, run the following command:

```
python3.12 ch01/abstractclass.py
```

You should get the following output:

```
Doing something with: 'some param'
```

Now you know how to define an interface and a concrete class implementing that interface in Python.

Protocols

Introduced in Python 3.8 via the `typing` module, *Protocols* offer a more flexible approach than ABCs, known as *structural duck typing*, where an object is considered valid if it has certain attributes or methods, regardless of its actual inheritance.

Unlike traditional duck typing, where type compatibility is determined at runtime, structural duck typing allows for type checking at compile time. This means that you can catch type errors before your code even runs (while in your IDE, for example), making your programs more robust and easier to debug.

The key advantage of using *Protocols* is that they focus on what an object can do, rather than what it is. In other words, *if an object walks like a duck and quacks like a duck, it's a duck*, regardless of its actual inheritance hierarchy. This is particularly useful in a dynamically typed language such as Python, where an object's behavior is more important than its actual type.

For example, you can define a `Drawable` protocol that requires a `draw()` method. Any class that implements this method would implicitly satisfy the protocol without having to explicitly inherit from it.

Here's a quick example to illustrate the concept. Let's say you need a Protocol named `Flyer` that requires a `fly()` method. You can define it as follows:

```
from typing import Protocol

class Flyer(Protocol):
    def fly(self) -> None:
        ...
```

And that's it! Now, any class that has a `fly()` method would be considered `Flyer`, whether it explicitly inherits from the `Flyer` class or not. This is a powerful feature that allows you to write more generic and reusable code and adheres to the principle of composition over inheritance, a principle that we previously discussed in the *Following the "Favor Composition Over Inheritance" principle* section.

In a later example, we will see a practical use of Protocols.

An example – different types of logger

Using ABCs, let's create a logging interface that allows for different types of logging mechanisms. Here's how you could implement that:

1. Import what is needed from abc:

    ```
    from abc import ABC, abstractmethod
    ```

2. Define the `Logger` interface with a `log` method:

```python
class Logger(ABC):
    @abstractmethod
    def log(self, message: str):
        pass
```

3. Then, define two concrete classes that implement the `Logger` interface for two different types of `Logger`:

```python
class ConsoleLogger(Logger):
    def log(self, message: str):
        print(f"Console: {message}")


class FileLogger(Logger):
    def log(self, message: str):
        with open("log.txt", "a") as f:
            f.write(f"File: {message}\n")
```

4. Next, to use each type of logger, define a function as follows:

```python
def log_message(logger: Logger, message: str):
    logger.log(message)
```

Notice that the function takes as its first argument an object of type `Logger`, meaning an instance of a concrete class that implements the `Logger` interface (i.e., `ConsoleLogger` or `FileLogger`).

5. Finally, add the lines needed to test the code, calling the `log_message` function as follows:

```python
if __name__ == "__main__":
    log_message(ConsoleLogger(), "A console log.")
    log_message(FileLogger(), "A file log.")
```

The complete code for this example (`ch01/interfaces.py`) is as follows:

```python
from abc import ABC, abstractmethod


class Logger(ABC):
    @abstractmethod
    def log(self, message: str):
        pass


class ConsoleLogger(Logger):
    def log(self, message: str):
```

```
                print(f"Console: {message}")


class FileLogger(Logger):
    def log(self, message: str):
        with open("log.txt", "a") as f:
            f.write(f"File: {message}\n")


def log_message(logger: Logger, message: str):
    logger.log(message)


if __name__ == "__main__":
    log_message(ConsoleLogger(), "A console log.")
    log_message(FileLogger(), "A file log.")
```

To test the code, run the following command:

```
python3.12 ch01/interfaces.py
```

You should get the following output:

```
Console: A console log.
```

In addition to that output, looking in the folder from which you run the command, you will find that a file called log.txt has been created, containing the following line:

```
File: A file log.
```

As you just saw with the log_message function, you can easily switch between different logging mechanisms without changing the function itself.

An example – different types of logger, now using Protocols

Let's revisit the previous example with the *Protocols* way of defining interfaces:

1. First, we need to import the Protocol class as follows:

    ```
    from typing import Protocol
    ```

2. Then, defining the Logger interface is done by inheriting from the Protocol class as follows:

```
class Logger(Protocol):
    def log(self, message: str):
        ...
```

And the rest of the code stays unchanged.

So, the complete code (ch01/interfaces_bis.py) is as follows:

```
from typing import Protocol


class Logger(Protocol):
    def log(self, message: str):
        ...


class ConsoleLogger:
    def log(self, message: str):
        print(f"Console: {message}")


class FileLogger:
    def log(self, message: str):
        with open("log.txt", "a") as f:
            f.write(f"File: {message}\n")


def log_message(logger: Logger, message: str):
    logger.log(message)


if __name__ == "__main__":
    log_message(ConsoleLogger(), "A console log.")
    log_message(FileLogger(), "A file log.")
```

To check the static typing of the code based on the protocol we defined, run the following command:

```
mypy ch01/interfaces_bis.py
```

You should get the following output:

```
Success: no issues found in 1 source file
```

To test the code, run the following command:

```
python3.12 ch01/interfaces_bis.py
```

You should get the same result as when running the previous version—in other words, the `log.txt` file created and the following output in the shell:

```
Console: A console log.
```

This is normal since the only thing we changed is the way we define the interface. And, the effect of the interface (the protocol) is not enforced at runtime, meaning it does not change the actual result of the code execution.

Following the Loose Coupling principle

As software grows in complexity, the relationships between its components can become tangled, leading to a system that is hard to understand, maintain, and extend. The principle of *Loose Coupling* aims to mitigate this issue.

What does it mean?

Loose coupling refers to minimizing the dependencies between different parts of a program. In a loosely coupled system, components are independent and interact through well-defined interfaces, making it easier to make changes to one part without affecting others.

Benefits

Loose coupling offers several advantages:

- **Maintainability**: With fewer dependencies, it's easier to update or replace individual components
- **Extensibility**: A loosely coupled system can be more easily extended with new features or components
- **Testability**: Independent components are easier to test in isolation, improving the overall quality of your software

Techniques for loose coupling

Two primary techniques for achieving *loose coupling* are **dependency injection** and the **observer pattern**. *Dependency injection* allows a component to receive its dependencies from an external source rather than creating them, making it easier to swap or mock these dependencies. The *observer pattern*, on the other hand, allows an object to publish changes to its state so that other objects can react accordingly, without being tightly bound to each other.

Both techniques aim to reduce the interdependencies between components, making the system you are building more modular and easier to manage.

We will discuss the *observer pattern* in detail in *Chapter 5, Behavioral Design Patterns*. For now, let's study an example to understand how to use the *dependency injection* technique.

An example – a message service

In Python, you can achieve loose coupling by using *dependency injection*. Let's see a simple example involving a MessageService class:

1. First, we define the MessageService class as follows:

    ```
    class MessageService:
        def __init__(self, sender):
            self.sender = sender

        def send_message(self, message):
            self.sender.send(message)
    ```

    As you can see, the class will be initialized by passing a sender object to it; that object has a send method to allow sending messages.

2. Second, let's define an EmailSender class:

    ```
    class EmailSender:
        def send(self, message):
            print(f"Sending email: {message}")
    ```

3. Third, let's define an SMSSender class:

    ```
    class SMSSender:
        def send(self, message):
            print(f"Sending SMS: {message}")
    ```

4. Now we can instantiate MessageService using an EmailSender object and use it to send a message. We can also instantiate MessageService using an SMSSender object instead. We add code to test both actions as follows:

    ```
    if __name__ == "__main__":
        email_service = MessageService(EmailSender())
        email_service.send_message("Hello via Email")

        sms_service = MessageService(SMSSender())
        sms_service.send_message("Hello via SMS")
    ```

The complete code for this example, saved in the ch01/loose_coupling.py file, is as follows:

```python
class MessageService:
    def __init__(self, sender):
        self.sender = sender

    def send_message(self, message: str):
        self.sender.send(message)


class EmailSender:
    def send(self, message: str):
        print(f"Sending email: {message}")


class SMSSender:
    def send(self, message: str):
        print(f"Sending SMS: {message}")


if __name__ == "__main__":
    email_service = MessageService(EmailSender())
    email_service.send_message("Hello via Email")

    sms_service = MessageService(SMSSender())
    sms_service.send_message("Hello via SMS")
```

To test the code, run the following command:

```
python3.12 ch01/loose_coupling.py
```

You should get the following output:

```
Sending email: Hello via Email
Sending SMS: Hello via SMS
```

In this example, MessageService is loosely coupled with EmailSender and SMSSender through dependency injection. This allows you to easily switch between different sending mechanisms without modifying the MessageService class.

Summary

We began the book with the foundational design principles that developers should follow for writing maintainable, flexible, and robust software. From encapsulating what varies to favoring composition, programming to interfaces, and aiming for loose coupling, these principles provide a strong foundation for any Python developer.

As you've seen, these principles are not just theoretical constructs but practical guidelines that can significantly improve the quality of your code. They set the stage for what comes next: diving deeper into more specialized sets of principles that guide object-oriented design.

In the next chapter, we will delve into the SOLID principles, a set of five design principles aimed at making software designs more understandable, flexible, and maintainable.

Summary

2
SOLID Principles

In the world of software engineering, principles and best practices are the backbone of a robust, maintainable, and efficient code base. In the previous chapter, we introduced the foundational principles every developer needs to follow.

In this chapter, we continue exploring design principles, focusing on **SOLID**, an acronym coined by Robert C. Martin, representing a set of five design principles he proposed, aimed at making software more understandable, flexible, and maintainable.

In this chapter, we're going to cover the following main topics:

- **Single responsibility principle (SRP)**
- **Open-closed principle (OCP)**
- **Liskov substitution principle (LSP)**
- **Interface segregation principle (ISP)**
- **Dependency inversion principle (DIP)**

By the end of this chapter, you'll have an understanding of these five additional design principles and how to apply them in Python.

Technical requirements

See the requirements presented in *Chapter 1*.

SRP

The SRP is a fundamental concept in software design. It advocates that when defining a class to provide functionality, that class should have only one reason to exist and should be responsible for only one aspect of the functionality. In simpler terms, it promotes the idea that each class should have one job or responsibility, and that job should be encapsulated within that class.

Thus by adhering to the SRP, you are essentially striving for classes that are focused, cohesive, and specialized in their functionality. This approach plays a crucial role in enhancing the maintainability and comprehensibility of your code base. When each class has a well-defined and single purpose, it becomes easier to manage, understand, and extend your code.

Of course, there is no obligation for you to follow the SRP. But knowing about the principle and thinking about your code with that in mind will improve your code base over time.

In practice, applying the SRP often leads to smaller, more focused classes, which can be combined and composed to create complex systems while maintaining a clear and organized structure.

> **Note**
>
> The SRP is not about minimizing the number of lines of code in a class but rather about ensuring that a class has a single reason to change, reducing the likelihood of unintended side effects when making modifications.

Let's go through a small example to make things more clear.

An example of software design following the SRP

Let's imagine some code that you could have in many different types of applications such as content or document management tools or a specialized web app, which includes functionality to generate a PDF file and save it to disk. To help understand the SRP, let's consider an initial version where the code does not follow this principle. In such a version, the developer would probably define a class dealing with reports, called `Report`, and would implement it in a way that makes it responsible for generating a report and also saving it to a file. The typical code for this class would look like the following:

```python
class Report:
    def __init__(self, content):
        self.content = content

    def generate(self):
        print(f"Report content: {self.content}")

    def save_to_file(self, filename):
        with open(filename, 'w') as file:
            file.write(self.content)
```

As you can see, the `Report` class has two responsibilities. First, generating a report, and then, saving the report's content to a file.

Of course, that is fine. But design principles encourage us to think about improving things for the future, as the requirements evolve and the code grows to handle complexity and change. Here, the

SRP teaches us to separate things. To adhere to the SRP, we can refactor that code to use two different classes that would each have one responsibility, as follows:

1. Create the first class, responsible for generating the report's content:

```
class Report:
    def __init__(self, content: str):
        self.content: str = content

    def generate(self):
        print(f"Report content: {self.content}")
```

2. Create a second class to deal with the need to save the report to a file:

```
class ReportSaver:
    def __init__(self, report: Report):
        self.report: Report = report

    def save_to_file(self, filename: str):
        with open(filename, 'w') as file:
            file.write(self.report.content)
```

3. To confirm that our refactored version works, let's add the following code to make it possible to immediately test things:

```
if __name__ == "__main__":
    report_content = "This is the content."
    report = Report(report_content)

    report.generate()

    report_saver = ReportSaver(report)
    report_saver.save_to_file("report.txt")
```

To recapitulate, here is the complete code, saved in the ch02/srp.py file:

```
class Report:
    def __init__(self, content: str):
        self.content: str = content

    def generate(self):
        print(f"Report content: {self.content}")


class ReportSaver:
    def __init__(self, report: Report):
```

```
        self.report: Report = report

    def save_to_file(self, filename: str):
        with open(filename, "w") as file:
            file.write(self.report.content)


if __name__ == "__main__":
    report_content = "This is the content."
    report = Report(report_content)

    report.generate()

    report_saver = ReportSaver(report)
    report_saver.save_to_file("report.txt")
```

To see the result of the code, run the following command:

```
python ch02/srp.py
```

You will get the following output:

```
Report content: This is the content.
```

In addition to that output, you will notice that a `report.txt` file has been created. So, everything works as expected.

As you can see, by following the SRP, you can achieve cleaner, more maintainable, and adaptable code, which contributes to the overall quality and longevity of your software projects.

OCP

The OCP is another fundamental principle in software design. It emphasizes that software entities, such as classes and modules, should be open for extension but closed for modification. What does that mean? It means that once a software entity is defined and implemented, it should not be changed to add new functionality. Instead, the entity should be extended through inheritance or interfaces to accommodate new requirements and behaviors.

When thinking about this principle and if you have some experience writing code for non-trivial programs, you can see how it makes sense, since modifying an entity introduces a risk of breaking some other part of the code base relying on it.

The OCP provides a robust foundation for building flexible and maintainable software systems. It allows developers to introduce new features or behaviors without altering the existing code base. By adhering to the OCP, you can minimize the risk of introducing bugs or unintended side effects when making changes to your software.

An example of design following the OCP

Consider a `Rectangle` class defined for rectangle shapes. Let's say we add a way to calculate the area of different shapes, maybe by using a function. The hypothetical code for the definition of both the class and the function could look like the following:

```
class Rectangle:
    def __init__(self, width:float, height: float):
        self.width: float = width
        self.height: float = height

def calculate_area(shape) -> float:
    if isinstance(shape, Rectangle):
        return shape.width * shape.height
```

> **Note**
>
> This code is not in the example code files. It is a hypothetical idea to start with in our thinking, and not the code you would end up using. Keep reading.

Given that code, if we want to add more shapes, we have to modify the `calculate_area` function. That is not ideal as we will keep coming back to change that code and that means more time testing things to avoid bugs.

As we aim to become good at writing maintainable code, let's see how we could improve that code by adhering to the OCP, while extending it to support another type of shape, the circle (using a `Circle` class):

1. Start by importing what we will need:

    ```
    import math
    from typing import Protocol
    ```

2. Define a `Shape` protocol for an interface providing a method for the shape's area:

    ```
    class Shape(Protocol):
        def area(self) -> float:
            ...
    ```

> **Note**
>
> Refer to *Chapter 1, Foundational Design Principles*, to understand Python's Protocol concept and technique.

3. Define the Rectangle class, which conforms to the Shape protocol:

```python
class Rectangle:
    def __init__(self, width: float, height: float):
        self.width: float = width
        self.height: float = height

    def area(self) -> float:
        return self.width * self.height
```

4. Also define the Circle class, which also conforms to the Shape protocol:

```python
class Circle:
    def __init__(self, radius: float):
        self.radius: float = radius

    def area(self) -> float:
        return math.pi * (self.radius**2)
```

5. Implement the calculate_area function in such a way that adding a new shape won't require us to modify it:

```python
def calculate_area(shape: Shape) -> float:
    return shape.area()
```

6. Add some code for testing the calculate_area function on the two types of shape objects:

```python
if __name__ == "__main__":
    rect = Rectangle(12, 8)
    rect_area = calculate_area(rect)
    print(f"Rectangle area: {rect_area}")

    circ = Circle(6.5)
    circ_area = calculate_area(circ)
    print(f"Circle area: {circ_area:.2f}")
```

The following is the complete code for this example, saved in the ch02/ocp.py file:

```python
import math
from typing import Protocol


class Shape(Protocol):
```

```python
    def area(self) -> float:
        ...


class Rectangle:
    def __init__(self, width: float, height: float):
        self.width: float = width
        self.height: float = height

    def area(self) -> float:
        return self.width * self.height


class Circle:
    def __init__(self, radius: float):
        self.radius: float = radius

    def area(self) -> float:
        return math.pi * (self.radius**2)


def calculate_area(shape: Shape) -> float:
    return shape.area()


if __name__ == "__main__":
    rect = Rectangle(12, 8)
    rect_area = calculate_area(rect)
    print(f"Rectangle area: {rect_area}")

    circ = Circle(6.5)
    circ_area = calculate_area(circ)
    print(f"Circle area: {circ_area:.2f}")
```

To see the result of this code, run the following command:

```
python ch02/ocp.py
```

You should get the following output:

```
Rectangle area: 96
Circle area: 132.73
```

Things work fine! The main win is that we were able to define a new shape without modifying the `calculate_area` function. The new design is elegant and allows ease of maintenance thanks to following the OCP.

So, you have now discovered another principle you should be using daily, which promotes designs both adaptable to evolving requirements and stable for their existing functionalities.

LSP

The LSP is another fundamental concept in object-oriented programming. It dictates how subclasses should relate to their superclasses. According to the LSP, if a program uses objects of a superclass, then the substitution of these objects with objects of a subclass should not change the correctness and expected behavior of the program.

Following this principle is important for maintaining the robustness of a software system. It ensures that, when using inheritance, subclasses extend their parent classes without altering their external behavior. For example, if a function works correctly with an object of a superclass, it should also work correctly with objects of any subclass of this superclass.

The LSP allows developers to introduce new subclass types without the risk of breaking existing functionality. This is particularly important in large-scale systems where changes in one part can have effects on other parts of the system. By following the LSP, developers can safely modify and extend classes, knowing that their new subclasses will integrate seamlessly with the established hierarchy and functionality.

An example of design following the LSP

Let's consider a `Bird` class and a `Penguin` class that subclasses it:

```
class Bird:
    def fly(self):
        print("I can fly")


class Penguin(Bird):
    def fly(self):
        print("I can't fly")
```

Then, for the needs of a hypothetical program that makes birds fly, we add a `make_bird_fly` function:

```
def make_bird_fly(bird):
    bird.fly()
```

With the current code, we can see that if we pass an instance of the Bird class to the function, we get the expected behavior (the bird will fly), whereas if we pass an instance of the Penguin class, we will get another behavior (it will not fly). You can analyze the code representing this first design provided in the ch02/lsp_violation.py file and run it to test this result. This shows us or at least gives us the intuition of what the LSP wants to help us avoid. So now, how could we improve the design by following the LSP?

To adhere to the LSP, we can refactor the code and introduce new classes to ensure that the behavior remains consistent:

1. We keep the Bird class, but we use a better method to represent the behavior we want; let's call it move(). The class will now look as follows:

    ```python
    class Bird:
        def move(self):
            print("I'm moving")
    ```

2. Then, we introduce a FlyingBird class and a FlightlessBird class, both inheriting from the Bird class:

    ```python
    class FlyingBird(Bird):
        def move(self):
            print("I'm flying")


    class FlightlessBird(Bird):
        def move(self):
            print("I'm walking")
    ```

3. Now, the make_bird_move function can be defined as follows:

    ```python
    def make_bird_move(bird):
        bird.move()
    ```

4. As usual, we add some code necessary to test the design:

    ```python
    if __name__ == "__main__":
        generic_bird = Bird()
        eagle = FlyingBird()
        penguin = FlightlessBird()

        make_bird_move(generic_bird)
        make_bird_move(eagle)
        make_bird_move(penguin)
    ```

The complete code for this new design, saved in the ch02/lsp.py file, is as follows:

```
class Bird:
    def move(self):
        print("I'm moving")


class FlyingBird(Bird):
    def move(self):
        print("I'm flying")


class FlightlessBird(Bird):
    def move(self):
        print("I'm walking")


def make_bird_move(bird):
    bird.move()


if __name__ == "__main__":
    generic_bird = Bird()
    eagle = FlyingBird()
    penguin = FlightlessBird()

    make_bird_move(generic_bird)
    make_bird_move(eagle)
    make_bird_move(penguin)
```

To test the example, run the following command:

```
python ch02/lsp.py
```

You should get the following output:

```
I'm moving
I'm flying
I'm walking
```

This output confirms the result we wanted to get in terms of design, which is maintaining the program's correctness when substituting a generic Bird class with a Penguin class or with an Eagle class; that is, each object moves whether it is an instance of a Bird class or an instance of a subclass. And that result was possible thanks to following the LSP.

This example demonstrates that all subclasses (`FlyingBird` and `FlightlessBird`) can be used in place of their superclass (`Bird`) without disrupting the expected behavior of the program. This conforms to the LSP.

ISP

The ISP advocates for designing smaller, more specific interfaces rather than broad, general-purpose ones. This principle states that a class should not be forced to implement interfaces it does not use. In the context of Python, this implies that a class shouldn't be forced to inherit and implement methods that are irrelevant to its purpose.

The ISP suggests that when designing software, one should avoid creating large, monolithic interfaces. Instead, the focus should be on creating smaller, more focused interfaces. This allows classes to only inherit or implement what they need, ensuring that each class only contains relevant and necessary methods.

Following this principle helps us build software with modularity, code readability and maintainability qualities, reduced side effects, and software that benefits from easier refactoring and testing, among other things.

An example of design following the ISP

Let's consider an `AllInOnePrinter` class that implements functionalities for printing, scanning, and faxing documents. The definition for that class would look like the following:

```
class AllInOnePrinter:
    def print_document(self):
        print("Printing")

    def scan_document(self):
        print("Scanning")

    def fax_document(self):
        print("Faxing")
```

If we wanted to introduce a specialized `SimplePrinter` class that only prints, it would have to implement or inherit the `scan_document` and `fax_document` methods (even though it only prints). That is not ideal.

To adhere to the ISP, we can create a separate interface for each functionality so that each class implements only the interfaces it needs.

> **Note about interfaces**
>
> Refer to the presentation in *Chapter 1*, *Foundational Design Principles*, of the **program to interfaces, not implementations principle**, to understand the importance of interfaces and the techniques we use in Python to define them (abstract base classes, protocols, etc.). In particular, here is the situation where protocols are the natural answer, that is, they help define small interfaces where each interface is created for doing only one thing.

1. Let's start by defining the three interfaces:

```python
from typing import Protocol


class Printer(Protocol):
    def print_document(self):
        ...


class Scanner(Protocol):
    def scan_document(self):
        ...


class Fax(Protocol):
    def fax_document(self):
        ...
```

2. Then, we keep the `AllInOnePrinter` class, which already implements the interfaces:

```python
class AllInOnePrinter:
    def print_document(self):
        print("Printing")

    def scan_document(self):
        print("Scanning")

    def fax_document(self):
        print("Faxing")
```

3. We add the `SimplePrinter` class, implementing the `Printer` interface, as follows:

```python
class SimplePrinter:
    def print_document(self):
        print("Simply Printing")
```

4. We also add a function that, when passed an object that implements the `Printer` interface, calls the right method on it to do the printing:

```python
def do_the_print(printer: Printer):
    printer.print_document()
```

5. Finally, we add code for testing the classes and the implemented interfaces:

```python
if __name__ == "__main__":
    all_in_one = AllInOnePrinter()
    all_in_one.scan_document()
    all_in_one.fax_document()
    do_the_print(all_in_one)

    simple = SimplePrinter()
    do_the_print(simple)
```

Here is the complete code for this new design (ch02/isp.py):

```python
from typing import Protocol

class Printer(Protocol):
    def print_document(self):
        ...


class Scanner(Protocol):
    def scan_document(self):
        ...


class Fax(Protocol):
    def fax_document(self):
        ...


class AllInOnePrinter:
    def print_document(self):
        print("Printing")

    def scan_document(self):
        print("Scanning")

    def fax_document(self):
```

```
        print("Faxing")


class SimplePrinter:
    def print_document(self):
        print("Simply Printing")


def do_the_print(printer: Printer):
    printer.print_document()


if __name__ == "__main__":
    all_in_one = AllInOnePrinter()
    all_in_one.scan_document()
    all_in_one.fax_document()
    do_the_print(all_in_one)

    simple = SimplePrinter()
    do_the_print(simple)
```

To test this code, run the following command:

```
python ch02/isp.py
```

You will get the following output:

```
Scanning
Faxing
Printing
Simply Printing
```

Because of the new design, each class only needs to implement the methods relevant to its behavior. This illustrates the ISP.

DIP

The DIP advocates that high-level modules should not depend directly on low-level modules. Instead, both should depend on abstractions or interfaces. By doing so, you decouple the high-level components from the details of the low-level components.

This principle allows for the reduction of the coupling between different parts of the system you are building, making it more maintainable and extendable, as we will see in an example.

Following the DIP brings loose coupling within a system because it encourages the use of interfaces as intermediaries between different parts of the system. When high-level modules depend on interfaces, they remain isolated from the specific implementations of low-level modules. This separation of concerns enhances maintainability and extensibility.

In essence, the DIP is closely linked to the loose coupling principle, which was covered in *Chapter 1, Foundational Design Principles*, by promoting a design where components interact through interfaces rather than concrete implementations. This reduces the interdependencies between modules, making it easier to modify or extend one part of the system without affecting others.

An example of design following the ISP

Consider a Notification class responsible for sending notifications via email, using an Email class. The code for both classes would look like the following:

```
class Email:
    def send_email(self, message):
        print(f"Sending email: {message}")


class Notification:
    def __init__(self):
        self.email = Email()

    def send(self, message):
        self.email.send_email(message)
```

> **Note about the code**
> This is not yet the final version of the example.

Currently, the high-level Notification class is dependent on the low-level Email class, and that is not ideal. To adhere to the DIP, we can introduce an abstraction, with a new code, as follows:

1. Define a MessageSender interface:

    ```
    from typing import Protocol


    class MessageSender(Protocol):
        def send(self, message: str):
            ...
    ```

2. Define the Email class, which implements the MessageSender interface, as follows:

    ```
    class Email:
        def send(self, message: str):
            print(f"Sending email: {message}")
    ```

3. Define the Notification class, which also implements the MessageSender interface, and has an object that implements MessageSender stored in its sender attribute, for handling the actual message sending. The code for that definition is as follows:

```python
class Notification:
    def __init__(self, sender: MessageSender):
        self.sender: MessageSender = sender

    def send(self, message: str):
        self.sender.send(message)
```

4. Finally, add some code for testing the design:

```python
if __name__ == "__main__":
    email = Email()
    notif = Notification(sender=email)
    notif.send(message="This is the message.")
```

The complete code for the implementation we just proposed is as follows (ch02/dip.py):

```python
from typing import Protocol


class MessageSender(Protocol):
    def send(self, message: str):
        ...


class Email:
    def send(self, message: str):
        print(f"Sending email: {message}")


class Notification:
    def __init__(self, sender: MessageSender):
        self.sender = sender

    def send(self, message: str):
        self.sender.send(message)


if __name__ == "__main__":
    email = Email()
    notif = Notification(sender=email)
    notif.send(message="This is the message.")
```

To test the code, run the following command:

```
python ch02/dip.py
```

You should get the following output:

```
Sending email: This is the message.
```

As you see, with the updated design, both `Notification` and `Email` are based on the `MessageSender` abstraction, so this design adheres to the DIP.

Summary

In this chapter, we explored additional principles to the ones presented in *Chapter 1, Foundational Design Principles*. Understanding and applying SOLID is crucial for writing maintainable, robust, and scalable Python code. These principles provide a strong foundation for good software design, making it easier to manage complexity, reduce errors, and improve the overall quality of your code.

In the next chapter, we will start exploring design patterns in Python, another essential topic for Python developers aiming for excellence.

Part 2:
From the Gang of Four

This part explores the classic design patterns from the Gang of Four (GoF), which are used to solve everyday problems, and how to apply them as a Python developer. This part includes the following chapters:

- *Chapter 3, Creational Design Patterns*
- *Chapter 4, Structural Design Patterns*
- *Chapter 5, Behavioral Design Patterns*

3

Creational Design Patterns

Design patterns are reusable programming solutions that have been used in various real-world contexts and have proved to produce expected results. They are shared among programmers and continue to be improved over time. This topic is popular thanks to the book by Erich Gamma, Richard Helm, Ralph Johnson, and John Vlissides, titled *Design Patterns: Elements of Reusable Object-Oriented Software*.

Here is a quote about design patterns from the *Gang of Four* book:

A design pattern systematically names, motivates, and explains a general design that addresses a recurring design problem in object-oriented systems. It describes the problem, the solution, when to apply the solution, and its consequences. It also gives implementation hints and examples. The solution is a general arrangement of objects and classes that solve the problem. The solution is customized and implemented to solve the problem in a particular context.

There are several categories of design patterns used in **object-oriented programming** (**OOP**), depending on the type of problem they address and/or the types of solutions they help us build. In their book, the *Gang of Four* presents 23 design patterns, split into three categories: *creational*, *structural*, and *behavioral*.

Creational design patterns are the first category we will cover throughout this chapter. These patterns deal with different aspects of object creation. Their goal is to provide better alternatives for situations where direct object creation, which in Python happens within the __init__() function, is not convenient.

In this chapter, we're going to cover the following main topics:

- The factory pattern
- The builder pattern
- The prototype pattern
- The singleton pattern
- The object pool pattern

By the end of the chapter, you will have a solid understanding of creational design patterns, whether they are useful or not in Python, and how to use them when they are useful.

Technical requirements

See the requirements presented in *Chapter 1*.

The factory pattern

We will start with the first creational design pattern from the *Gang of Four* book: the factory design pattern. In the factory design pattern, a client (meaning client code) asks for an object without knowing where the object is coming from (that is, which class is used to generate it). The idea behind a factory is to simplify the object creation process. It is easier to track which objects are created if this is done through a central function, compared to letting a client create objects using a direct class instantiation. A factory reduces the complexity of maintaining an application by decoupling the code that creates an object from the code that uses it.

Factories typically come in two forms—the factory method, which is a method (or simply a function for a Python developer) that returns a different object per input parameter, and the abstract factory, which is a group of factory methods used to create a family of related objects.

Let's discuss the two forms of *factory pattern*, starting with the factory method.

The factory method

The factory method is based on a single function written to handle our object creation task. We execute it, passing a parameter that provides information about what we want, and, as a result, the wanted object is created.

Interestingly, when using the factory method, we are not required to know any details about how the resulting object is implemented and where it is coming from.

Real-world examples

We can find the factory method pattern used in real life in the context of a plastic toy construction kit. The molding material used to construct plastic toys is the same, but different toys (different figures or shapes) can be produced using the right plastic molds. This is like having a factory method in which the input is the name of the toy that we want (for example, a duck or car) and the output (after the molding) is the plastic toy that was requested.

In the software world, the Django web framework uses the factory method pattern for creating the fields of a web form. The `forms` module included in Django (`https://github.com/django/django/blob/main/django/forms/forms.py`) supports the creation of different kinds of fields (for example, `CharField`, `EmailField`, and so on). Parts of their behavior can be customized using attributes such as `max_length` and `required`.

Use cases for the factory method pattern

If you realize that you cannot track the objects created by your application because the code that creates them is in many different places instead of in a single function/method, you should consider using the factory method pattern. The factory method centralizes object creation and tracking your objects becomes much easier. Note that it is fine to create more than one factory method, and this is how it is typically done in practice. Each factory method logically groups the creation of objects that have similarities. For example, one factory method might be responsible for connecting you to different databases (MySQL, SQLite); another factory method might be responsible for creating the geometrical object that you request (circle, triangle); and so on.

The factory method is also useful when you want to decouple object creation from object usage. We are not coupled to a specific class when creating an object; we just provide partial information about what we want by calling a function. This means that introducing changes to the function is easy and does not require any changes to the code that uses it.

Another use case worth mentioning is related to improving the performance and memory usage of an application. A factory method can improve performance and memory usage by creating new objects only if it is necessary. When we create objects using a direct class instantiation, extra memory is allocated every time a new object is created (unless the class uses caching internally, which is usually not the case). We can see that in practice in the following code (ch03/factory/id.py), which creates two instances of the same class, MyClass, and uses the id() function to compare their memory addresses. The addresses are also printed in the output so that we can inspect them. The fact that the memory addresses are different means that two distinct objects are created. The code is as follows:

```
class MyClass:
    pass


if __name__ == "__main__":
    a = MyClass()
    b = MyClass()

    print(id(a) == id(b))
    print(id(a))
    print(id(b))
```

Executing the code (ch03/factory/id.py) on my computer results in the following output:

```
False
4330224656
4331646704
```

> **Note**
>
> The addresses that you see if you execute the file, where the `id()` function is called, are not the same as the ones I see because they depend on the current memory layout and allocation. But the result must be the same—the two addresses should be different. There's one exception that happens if you write and execute the code in the Python **Read-Eval-Print Loop** (**REPL**)—or, simply put, the interactive prompt—but that's a REPL-specific optimization that does not happen normally.

Implementing the factory method pattern

Data comes in many forms. There are two main file categories for storing/retrieving data: human-readable files and binary files. Examples of human-readable files are XML, RSS/Atom, YAML, and JSON. Examples of binary files are the `.sq3` file format used by SQLite and the `.mp3` audio file format used to listen to music.

In this example, we will focus on two popular human-readable formats—XML and JSON. Although human-readable files are generally slower to parse than binary files, they make data exchange, inspection, and modification much easier. For this reason, it is advised that you work with human-readable files unless there are other restrictions that do not allow it (mainly unacceptable performance or proprietary binary formats).

In this case, we have some input data stored in an XML and a JSON file, and we want to parse them and retrieve some information. At the same time, we want to centralize the client's connection to those (and all future) external services. We will use the factory method to solve this problem. The example focuses only on XML and JSON, but adding support for more services should be straightforward.

First, let's look at the data files.

The JSON file, `movies.json`, is a sample of a dataset containing information about American movies (title, year, director name, genre, and so on):

```
[
  {
    "title": "After Dark in Central Park",
    "year": 1900,
    "director": null,
    "cast": null,
    "genre": null
  },
  {
    "title": "Boarding School Girls' Pajama Parade",
    "year": 1900,
    "director": null,
    "cast": null,
```

```
      "genre": null
  },
  {

      "title": "Buffalo Bill's Wild West Parad",
      "year": 1900,
      "director": null,
      "cast": null,
      "genre": null
  },
  {

      "title": "Caught",
      "year": 1900,
      "director": null,
      "cast": null,
      "genre": null
  },
  {

      "title": "Clowns Spinning Hats",
      "year": 1900,
      "director": null,
      "cast": null,
      "genre": null
  },
  {

      "title": "Capture of Boer Battery by British",
      "year": 1900,
      "director": "James H. White",
      "cast": null,
      "genre": "Short documentary"
  },
  {

      "title": "The Enchanted Drawing",
      "year": 1900,
      "director": "J. Stuart Blackton",
      "cast": null,
      "genre": null
  },
  {

      "title": "Family Troubles",
      "year": 1900,
      "director": null,
      "cast": null,
      "genre": null
```

```
    },
    {
      "title": "Feeding Sea Lions",
      "year": 1900,
      "director": null,
      "cast": "Paul Boyton",
      "genre": null
    }
]
```

The XML file, person.xml, contains information about individuals (firstName, lastName, gender, and so on), as follows:

1. We start with the enclosing tag of the persons XML container:

   ```
   <persons>
   ```

2. Then, an XML element representing a person's data code is presented as follows:

   ```
   <person>
     <firstName>John</firstName>
     <lastName>Smith</lastName>
     <age>25</age>
     <address>
       <streetAddress>21 2nd Street</streetAddress>
       <city>New York</city>
       <state>NY</state>
       <postalCode>10021</postalCode>
     </address>
     <phoneNumbers>
       <number type="home">212 555-1234</number>
       <number type="fax">646 555-4567</number>
     </phoneNumbers>
     <gender>
       <type>male</type>
     </gender>
   </person>
   ```

3. An XML element representing another person's data is shown by the following code:

   ```
   <person>
     <firstName>Jimy</firstName>
     <lastName>Liar</lastName>
     <age>19</age>
     <address>
   ```

```xml
    <streetAddress>18 2nd Street</streetAddress>
    <city>New York</city>
    <state>NY</state>
    <postalCode>10021</postalCode>
  </address>
  <phoneNumbers>
    <number type="home">212 555-1234</number>
  </phoneNumbers>
  <gender>
    <type>male</type>
  </gender>
</person>
```

4. An XML element representing a third person's data is shown by the following code:

```xml
<person>
  <firstName>Patty</firstName>
  <lastName>Liar</lastName>
  <age>20</age>
  <address>
    <streetAddress>18 2nd Street</streetAddress>
    <city>New York</city>
    <state>NY</state>
    <postalCode>10021</postalCode>
  </address>
  <phoneNumbers>
    <number type="home">212 555-1234</number>
    <number type="mobile">001 452-8819</number>
  </phoneNumbers>
  <gender>
    <type>female</type>
  </gender>
</person>
```

5. Finally, we close the XML container:

```xml
</persons>
```

We will use two libraries that are part of the Python distribution for working with JSON and XML: `json` and `xml.etree.ElementTree`.

We start by importing what we need for the various manipulations (json, ElementTree, and pathlib), and we define a JSONDataExtractor class, loading the data from the file and using the parsed_data property to get it. That part of the code is as follows:

```python
import json
import xml.etree.ElementTree as ET
from pathlib import Path

class JSONDataExtractor:
    def __init__(self, filepath: Path):
        self.data = {}
        with open(filepath) as f:
            self.data = json.load(f)

    @property
    def parsed_data(self):
        return self.data
```

We also define an XMLDataExtractor class, loading the data in the file via ElementTree's parser, and using the parsed_data property to get the result, as follows:

```python
class XMLDataExtractor:
    def __init__(self, filepath: Path):
        self.tree = ET.parse(filepath)

    @property
    def parsed_data(self):
        return self.tree
```

Now, we provide the factory function that helps select the right data extractor class depending on the target file's extension (or raise an exception if it is not supported), as follows:

```python
def extract_factory(filepath: Path):
    ext = filepath.name.split(".")[-1]
    if ext == "json":
        return JSONDataExtractor(filepath)
    elif ext == "xml":
        return XMLDataExtractor(filepath)
    else:
        raise ValueError("Cannot extract data")
```

Next, we define the main function of our program, `extract()`; in the first part of the function, the code handles the JSON case, as follows:

```python
def extract(case: str):
    dir_path = Path(__file__).parent

    if case == "json":
        path = dir_path / Path("movies.json")
        factory = extract_factory(path)
        data = factory.parsed_data

        for movie in data:
            print(f"- {movie['title']}")
            director = movie["director"]
            if director:
                print(f"    Director: {director}")
            genre = movie["genre"]
            if genre:
                print(f"    Genre: {genre}")
```

We add the final part of the `extract()` function, working with the XML file using the factory method. XPath is used to find all person elements that have the last name `Liar`. For each matched person, the basic name and phone number information are shown. The code is as follows:

```python
    elif case == "xml":
        path = dir_path / Path("person.xml")
        factory = extract_factory(path)
        data = factory.parsed_data

        search_xpath = ".//person[lastName='Liar']"
        items = data.findall(search_xpath)
        for item in items:
            first = item.find("firstName").text
            last = item.find("lastName").text
            print(f"- {first} {last}")
            for pn in item.find("phoneNumbers"):
                pn_type = pn.attrib["type"]
                pn_val = pn.text
                phone = f"{pn_type}: {pn_val}"
                print(f"    {phone}")
```

Finally, we add some testing code:

```python
if __name__ == "__main__":
    print("* JSON case *")
    extract(case="json")
    print("* XML case *")
    extract(case="xml")
```

Here is a summary of the implementation (in the ch03/factory/factory_method.py file):

1. After importing the modules we need, we start by defining a JSON data extractor class (JSONDataExtractor) and an XML data extractor class (XMLDataExtractor).

2. We add a factory function, extract_factory(), to get the right data extractor class to instantiate.

3. We also add our wrapper and main function, extract().

4. Finally, we add testing code, where we extract data from a JSON file and an XML file and parse the resulting text.

To test the example, run the following command:

```
python ch03/factory/factory_method.py
```

You should get the following output:

```
* JSON case *
- After Dark in Central Park
- Boarding School Girls' Pajama Parade
- Buffalo Bill's Wild West Parad
- Caught
- Clowns Spinning Hats
- Capture of Boer Battery by British
    Director: James H. White
    Genre: Short documentary
- The Enchanted Drawing
    Director: J. Stuart Blackton
- Family Troubles
- Feeding Sea Lions
* XML case *
- Jimy Liar
    home: 212 555-1234
- Patty Liar
    home: 212 555-1234
    mobile: 001 452-8819
```

Notice that although `JSONDataExtractor` and `XMLDataExtractor` have the same interfaces, what is returned by `parsed_data()` is not handled in a uniform way; in one case we have a list, and in the other, we have a tree. Different Python code must be used to work with each data extractor. Although it would be nice to be able to use the same code for all extractors, this is not realistic for the most part unless we use some kind of common mapping for the data, which is often provided by external data providers. Assuming that you can use the same code for handling the XML and JSON files, what changes are required to support a third format—for example, SQLite? Find an SQLite file or create your own and try it.

Should you use the factory method pattern?

The main critique that veteran Python developers often express toward the factory method pattern is that it can be considered over-engineered or unnecessarily complex for many use cases. Python's dynamic typing and first-class functions often allow for simpler, more straightforward solutions to problems that the factory method aims to solve. In Python, you can often use simple functions or class methods to create objects directly without needing to create separate factory classes or functions. This keeps the code more readable and Pythonic, adhering to the language's philosophy of *Simple is better than complex.*

Also, Python's support for default arguments, keyword arguments, and other language features often makes it easier to extend constructors in a backward-compatible way, reducing the need for separate factory methods. So, while the factory method pattern is a well-established design pattern in statically typed languages such as Java or C++, it is often seen as too cumbersome or verbose for Python's more flexible and dynamic nature.

To show how one could deal with simple use cases without the factory method pattern, an alternative implementation has been provided in the `ch03/factory/factory_method_not_needed.py` file. As you can see, there is no more factory. And the following extract from the code shows what we mean when we say that in Python, you just create objects where you need them, without an intermediary function or class, which makes your code more Pythonic:

```
if case == "json":
    path = dir_path / Path("movies.json")
    data = JSONDataExtractor(path).parsed_data
```

The abstract factory pattern

The abstract factory pattern is a generalization of the factory method idea. Basically, an abstract factory is a (logical) group of factory methods, where each factory method is responsible for generating a different kind of object.

We are going to discuss some examples, use cases, and a possible implementation.

Real-world examples

The abstract factory is used in car manufacturing. The same machinery is used for stamping the parts (doors, panels, hoods, fenders, and mirrors) of different car models. The model that is assembled by the machinery is configurable and easy to change at any time.

In the software category, the `factory_boy` package (`https://github.com/FactoryBoy/factory_boy`) provides an abstract factory implementation for creating Django models in tests. An alternative tool is `model_bakery` (`https://github.com/model-bakers/model_bakery`). Both packages are used for creating instances of models that support test-specific attributes. This is important because, this way, the readability of your tests is improved, and you avoid sharing unnecessary code.

> **Note**
>
> Django models are special classes used by the framework to help store and interact with data in the database (tables). See the Django documentation (`https://docs.djangoproject.com`) for more details.

Use cases for the abstract factory pattern

Since the abstract factory pattern is a generalization of the factory method pattern, it offers the same benefits: it makes tracking an object creation easier, it decouples object creation from object usage, and it gives us the potential to improve the memory usage and performance of our application.

Implementing the abstract factory pattern

To demonstrate the abstract factory pattern, I will reuse one of my favorite examples, included in the book *Python 3 Patterns, Recipes and Idioms*, by Bruce Eckel. Imagine that we are creating a game or we want to include a mini-game as part of our application to entertain our users. We want to include at least two games, one for children and one for adults. We will decide which game to create and launch at runtime, based on user input. An abstract factory takes care of the game creation part.

Let's start with the kids' game. It is called **FrogWorld**. The main hero is a frog who enjoys eating bugs. Every hero needs a good name, and in our case, the name is given by the user at runtime. The `interact_with()` method is used to describe the interaction of the frog with an obstacle (for example, a bug, puzzle, and other frogs) as follows:

```python
class Frog:
    def __init__(self, name):
        self.name = name

    def __str__(self):
        return self.name
```

```
    def interact_with(self, obstacle):
        act = obstacle.action()
        msg = f"{self} the Frog encounters {obstacle} and {act}!"
        print(msg)
```

There can be many kinds of obstacles, but for our example, an obstacle can only be a bug. When the frog encounters a bug, only one action is supported. It eats it:

```
class Bug:
    def __str__(self):
        return "a bug"

    def action(self):
        return "eats it"
```

The `FrogWorld` class is an abstract factory. Its main responsibilities are creating the main character and the obstacle(s) in the game. Keeping the creation methods separate and their names generic (for example, `make_character()` and `make_obstacle()`) allows us to change the active factory (and, therefore, the active game) dynamically without any code changes. The code is as follows:

```
class FrogWorld:
    def __init__(self, name):
        print(self)
        self.player_name = name

    def __str__(self):
        return "\n\n\t------ Frog World -------"

    def make_character(self):
        return Frog(self.player_name)

    def make_obstacle(self):
        return Bug()
```

The **WizardWorld** game is similar. The only difference is that the wizard battles against monsters such as orks instead of eating bugs!

Here is the definition of the `Wizard` class, which is similar to the `Frog` one:

```
class Wizard:
    def __init__(self, name):
        self.name = name

    def __str__(self):
        return self.name
```

```
    def interact_with(self, obstacle):
        act = obstacle.action()
        msg = f"{self} the Wizard battles against {obstacle} and
{act}!"
        print(msg)
```

Then, the definition of the `Ork` class is as follows:

```
class Ork:
    def __str__(self):
        return "an evil ork"

    def action(self):
        return "kills it"
```

We also need to define a `WizardWorld` class, similar to the `FrogWorld` one that we have discussed; the obstacle, in this case, is an `Ork` instance:

```
class WizardWorld:
    def __init__(self, name):
        print(self)
        self.player_name = name

    def __str__(self):
        return "\n\n\t------ Wizard World -------"

    def make_character(self):
        return Wizard(self.player_name)

    def make_obstacle(self):
        return Ork()
```

The `GameEnvironment` class is the main entry point of our game. It accepts the factory as an input and uses it to create the world of the game. The `play()` method initiates the interaction between the created hero and the obstacle, as follows:

```
class GameEnvironment:
    def __init__(self, factory):
        self.hero = factory.make_character()
        self.obstacle = factory.make_obstacle()

    def play(self):
        self.hero.interact_with(self.obstacle)
```

The `validate_age()` function prompts the user to give a valid age. If the age is not valid, it returns a tuple with the first element set to `False`. If the age is fine, the first element of the tuple is set to `True`, and that's the case where we care about the second element of the tuple, which is the age given by the user, as follows:

```
def validate_age(name):
    age = None
    try:
        age_input = input(
            f"Welcome {name}. How old are you? "
        )
        age = int(age_input)
    except ValueError:
        print(
            f"Age {age} is invalid, please try again..."
        )
        return False, age
    return True, age
```

Finally comes the `main()` function definition, followed by calling it. It asks for the user's name and age and decides which game should be played, given the age of the user, as follows:

```
def main():
    name = input("Hello. What's your name? ")
    valid_input = False
    while not valid_input:
        valid_input, age = validate_age(name)
    game = FrogWorld if age < 18 else WizardWorld
    environment = GameEnvironment(game(name))
    environment.play()

if __name__ == "__main__":
    main()
```

The summary for the implementation we just discussed (see the complete code in the `ch03/factory/abstract_factory.py` file) is as follows:

1. We define `Frog` and `Bug` classes for the **FrogWorld** game.

2. We add a `FrogWorld` class, where we use our `Frog` and `Bug` classes.

3. We define `Wizard` and `Ork` classes for the **WizardWorld** game.

4. We add a `WizardWorld` class, where we use our `Wizard` and `Ork` classes.

5. We define a `GameEnvironment` class.

6. We add a `validate_age()` function.

7. Finally, we have the `main()` function, followed by the conventional trick for calling it. The following are the aspects of this function:

    - We get the user's input for name and age.

    - We decide which game class to use based on the user's age.

    - We instantiate the right game class, and then the `GameEnvironment` class.

    - We call `.play()` on the `environment` object to play the game.

Let's call this program using the `python ch03/factory/abstract_factory.py` command and see some sample output.

The sample output for a teenager is as follows:

```
Hello. What's your name? Arthur
Welcome Arthur. How old are you? 13



------ Frog World -------
Arthur the Frog encounters a bug and eats it!
```

The sample output for an adult is as follows:

```
Hello. What's your name? Tom
Welcome Tom. How old are you? 34



------ Wizard World -------
Tom the Wizard battles against an evil ork and kills it!
```

Try extending the game to make it more complete. You can go as far as you want; create many obstacles, many enemies, and whatever else you like.

The builder pattern

We just covered the first two creational patterns, the factory method and the abstract factory, which both offer approaches to improve the way we create objects in nontrivial cases.

Now, imagine that we want to create an object that is composed of multiple parts, and the composition needs to be done step by step. The object is not complete unless all its parts are fully created. That's where the builder design pattern can help us. The builder pattern separates the construction of a complex object from its representation. By keeping the construction separate from the representation, the same construction can be used to create several different representations.

Real-world examples

In our everyday life, the builder design pattern is used in fast-food restaurants. The same procedure is always used to prepare a burger and the packaging (box and paper bag), even if there are many kinds of burgers (classic, cheeseburger, and more) and different packages (small-sized box, medium-sized box, and so forth). The difference between a classic burger and a cheeseburger is in the representation and not in the construction procedure. In this case, the director is the cashier who gives instructions about what needs to be prepared to the crew, and the builder is the person from the crew who takes care of the specific order.

In software, we can think of the `django-query-builder` library (`https://github.com/ambitioninc/django-query-builder`), a third-party Django library that relies on the builder pattern. This library can be used for building SQL queries dynamically, allowing you to control all aspects of a query and create a different range of queries, from simple to very complex ones.

Comparison with the factory pattern

At this point, the distinction between the builder pattern and the factory pattern might not be very clear. The main difference is that a factory pattern creates an object in a single step, whereas a builder pattern creates an object in multiple steps and almost always uses a *director*.

Another difference is that while the factory pattern returns a created object immediately, in the builder pattern, the client code explicitly asks the director to return the final object when it needs it.

Use cases for the builder pattern

The builder pattern is particularly useful when an object needs to be constructed with numerous possible configurations. A typical case is a situation where a class has multiple constructors with a varying number of parameters, often leading to confusion or error-prone code.

The pattern is also beneficial when the object's construction process is more complex than simply setting initial values. For example, if an object's full creation involves multiple steps, such as parameter validation, setting up data structures, or even making calls to external services, the builder pattern can encapsulate this complexity.

Implementing the builder pattern

Let's see how we can use the builder design pattern to make a pizza-ordering application. This example is particularly interesting because a pizza is prepared in steps that should follow a specific order. To add the sauce, you first need to prepare the dough. To add the topping, you first need to add the sauce. And you can't start baking the pizza unless both the sauce and the topping are placed on the dough. Moreover, each pizza usually requires a different baking time, depending on the thickness of its dough and the topping used.

We start by importing the required modules and declaring a few Enum parameters plus a constant that is used many times in the application. The STEP_DELAY constant is used to add a time delay between the different steps of preparing a pizza (prepare the dough, add the sauce, and so on) as follows:

```
import time
from enum import Enum

PizzaProgress = Enum(
    "PizzaProgress", "queued preparation baking ready"
)
PizzaDough = Enum("PizzaDough", "thin thick")
PizzaSauce = Enum("PizzaSauce", "tomato creme_fraiche")
PizzaTopping = Enum(
    "PizzaTopping",
    "mozzarella double_mozzarella bacon ham mushrooms red_onion
oregano",
)
# Delay in seconds
STEP_DELAY = 3
```

Our end product is a pizza, which is described by the Pizza class. When using the builder pattern, the end product does not have many responsibilities, since it is not supposed to be instantiated directly. A builder creates an instance of the end product and makes sure that it is properly prepared. That's why the Pizza class is so minimal. It basically initializes all data to sane default values. An exception is the prepare_dough() method.

The prepare_dough() method is defined in the Pizza class instead of a builder for two reasons. First, to clarify the fact that the end product is typically minimal, which does not mean that you should never assign it any responsibilities. Second, to promote code reuse through composition.

So, we define our Pizza class as follows:

```
class Pizza:
    def __init__(self, name):
        self.name = name
        self.dough = None
        self.sauce = None
        self.topping = []

    def __str__(self):
        return self.name

    def prepare_dough(self, dough):
        self.dough = dough
        print(
```

```
            f"preparing the {self.dough.name} dough of your {self}..."
        )
        time.sleep(STEP_DELAY)
        print(f"done with the {self.dough.name} dough")
```

There are two builders: one for creating a margarita pizza (MargaritaBuilder) and another for creating a creamy bacon pizza (CreamyBaconBuilder). Each builder creates a Pizza instance and contains methods that follow the pizza-making procedure: prepare_dough(), add_sauce(), add_topping(), and bake(). To be precise, prepare_dough() is just a wrapper to the prepare_dough() method of the Pizza class.

Notice how each builder takes care of all the pizza-specific details. For example, the topping of the margarita pizza is double mozzarella and oregano, while the topping of the creamy bacon pizza is mozzarella, bacon, ham, mushrooms, red onion, and oregano.

An extract of the code of the MargaritaBuilder class is as follows (see the ch03/builder.py file for the whole code):

```python
class MargaritaBuilder:
    def __init__(self):
        self.pizza = Pizza("margarita")
        self.progress = PizzaProgress.queued
        self.baking_time = 5

    def prepare_dough(self):
        self.progress = PizzaProgress.preparation
        self.pizza.prepare_dough(PizzaDough.thin)

    ...
```

An extract of the code of the CreamyBaconBuilder class is as follows:

```python
class CreamyBaconBuilder:
    def __init__(self):
        self.pizza = Pizza("creamy bacon")
        self.progress = PizzaProgress.queued
        self.baking_time = 7

    def prepare_dough(self):
        self.progress = PizzaProgress.preparation
        self.pizza.prepare_dough(PizzaDough.thick)

    ...
```

The *director* in this example is the waiter. The core of the `Waiter` class is the `construct_pizza()` method, which accepts a builder as a parameter and executes all the pizza-preparation steps in the right order. Choosing the appropriate builder, which can even be done at runtime, gives us the ability to create different pizza styles without modifying any of the code of the director (`Waiter`). The `Waiter` class also contains the `pizza()` method, which returns the end product (prepared pizza) as a variable to the caller. The code for that class is as follows:

```
class Waiter:
    def __init__(self):
        self.builder = None

    def construct_pizza(self, builder):
        self.builder = builder
        steps = (
            builder.prepare_dough,
            builder.add_sauce,
            builder.add_topping,
            builder.bake,
        )
        [step() for step in steps]

    @property
    def pizza(self):
        return self.builder.pizza
```

The `validate_style()` method is similar to the `validate_age()` function, as described in the section titled *The factory pattern* earlier in this chapter. It is used to make sure that the user gives valid input, which in this case is a character that is mapped to a pizza builder. The m character uses the `MargaritaBuilder` class, and the c character uses the `CreamyBaconBuilder` class. These mappings are in the `builder` parameter. A tuple is returned, with the first element set to `True` if the input is valid or `False` if it is invalid, as follows:

```
def validate_style(builders):
    try:
        input_msg = "What pizza would you like, [m]argarita or [c]
reamy bacon? "
        pizza_style = input(input_msg)
        builder = builders[pizza_style]()
        valid_input = True
    except KeyError:
        error_msg = "Sorry, only margarita (key m) and creamy bacon
 (key c) are available"
        print(error_msg)
```

```
        return (False, None)
    return (True, builder)
```

The last part is the `main()` function. The `main()` function contains code for instantiating a pizza builder. The pizza builder is then used by the `Waiter` director to prepare the pizza. The created pizza can be delivered to the client at any later point:

```
def main():
    builders = dict(m=MargaritaBuilder, c=CreamyBaconBuilder)
    valid_input = False
    while not valid_input:
        valid_input, builder = validate_style(builders)
    print()
    waiter = Waiter()
    waiter.construct_pizza(builder)
    pizza = waiter.pizza
    print()
    print(f"Enjoy your {pizza}!")
```

Here is a summary of the implementation (in the `ch03/builder.py` file):

1. We start with a couple of imports we need, for the standard `Enum` class and `time` module.

2. We declare variables for a few constants: `PizzaProgress`, `PizzaDough`, `PizzaSauce`, `PizzaTopping`, and `STEP_DELAY`.

3. We define our `Pizza` class.

4. We define classes for two builders, `MargaritaBuilder` and `CreamyBaconBuilder`.

5. We define our `Waiter` class.

6. We add a `validate_style()` function to improve things regarding exception handling.

7. Finally, we have the `main()` function, followed by a snippet for calling it when the program is run. In the `main()` function, the following happens:

    -   We make it possible to choose the pizza builder based on the user's input, after validation via the `validate_style()` function.

    -   The pizza builder is used by the waiter for preparing the pizza.

    -   The created pizza is then delivered.

Here is the output produced by calling the `python ch03/builder.py` command to execute this example program:

```
What pizza would you like, [m]argarita or [c]reamy bacon? c

preparing the thick dough of your creamy bacon...
done with the thick dough
adding the crème fraîche sauce to your creamy bacon
done with the crème fraîche sauce
adding the topping (mozzarella, bacon, ham, mushrooms, red onion,
oregano) to your creamy bacon
done with the topping (mozzarella, bacon, ham, mushrooms, red onion,
oregano)
baking your creamy bacon for 7 seconds
your creamy bacon is ready

Enjoy your creamy bacon!
```

That was a nice result.

But... supporting only two pizza types is a shame. Feel like getting a Hawaiian pizza builder? Consider using *inheritance* after thinking about the advantages and disadvantages. Or *composition*, which has its advantages, as we have seen in *Chapter 1, Foundational Design Principles*.

The prototype pattern

The prototype pattern allows you to create new objects by copying existing ones, rather than creating them from scratch. This pattern is particularly useful when the cost of initializing an object is more expensive or complex than copying an existing one. In essence, the prototype pattern enables you to create a new instance of a class by duplicating an existing instance, thereby avoiding the overhead of initializing a new object.

In its simplest version, this pattern is just a `clone()` function that accepts an object as an input parameter and returns a clone of it. In Python, this can be done using the `copy.deepcopy()` function.

Real-world examples

Cloning a plant by taking a cutting is a real-world example of the prototype pattern. Using this approach, you don't grow the plant from a seed; you create a new plant that's a copy of an existing one.

Many Python applications make use of the prototype pattern, but it is rarely referred to as *prototype* since cloning objects is a built-in feature of the Python language.

Use cases for the prototype pattern

The prototype pattern is useful when we have an existing object that needs to stay untouched and we want to create an exact copy of it, allowing changes in some parts of the copy.

There is also the frequent need for duplicating an object that is populated from a database and has references to other database-based objects. It is costly (multiple queries to a database) to clone such a complex object, so a prototype is a convenient way to solve the problem.

Implementing the prototype pattern

Nowadays, some organizations, even of small size, deal with many websites and apps via their infrastructure/DevOps teams, hosting providers, or **cloud service providers (CSPs)**.

When you have to manage multiple websites, there is a point where it becomes difficult to follow. You need to access information quickly, such as IP addresses that are involved, domain names and their expiration dates, and maybe details about DNS parameters. So, you need a kind of inventory tool.

Let's imagine how these teams deal with this type of data for daily activities, and touch on the implementation of a piece of software that helps consolidate and maintain the data (other than in Excel spreadsheets).

First, we need to import Python's standard `copy` module, as follows:

```
import copy
```

At the heart of this system, we will have a `Website` class for holding all useful information such as the name, the domain name, a description, the author of a website we are managing, and so on.

In the `__init__()` method of the class, only some parameters are fixed: `name`, `domain`, and `description`. But we also want flexibility, and client code can pass more parameters in the form of keywords (`name=value`) using the `kwargs` variable-length collection (each pair becomes an item of the `kwargs` Python dictionary).

> **Additional information**
>
> There is a Python idiom that helps to set an arbitrary attribute named `attr` with a `val` value on an `obj` object, using the `setattr()` built-in function: `setattr(obj, attr, val)`.

So we are defining a `Website` class and initializing its objects, using the `setattr` technique for optional attributes, as follows:

```
class Website:
    def __init__(
        self,
        name: str,
```

```
        domain: str,
        description: str,
        **kwargs,
    ):
        self.name = name
        self.domain = domain
        self.description = description

        for key in kwargs:
            setattr(self, key, kwargs[key])
```

That's not all. To improve the usability of the class, we also add its string representation method (__str__()). We extract the values of all instance attributes, using the vars() trick, and inject those values into the string that the method returns. Also, since we plan to clone objects, we include the object's memory address using the id() function. The code is as follows:

```
def __str__(self) -> str:
    summary = [
        f"- {self.name} (ID: {id(self)})\n",
    ]

    infos = vars(self).items()
    ordered_infos = sorted(infos)
    for attr, val in ordered_infos:
        if attr == "name":
            continue
        summary.append(f"{attr}: {val}\n")

    return "".join(summary)
```

> **Additional information**
>
> The vars() function in Python returns the __dict__ attribute of an object. The __dict__ attribute is a dictionary containing the object's attributes (both data attributes and methods). This function is useful for debugging, as it allows you to inspect the attributes and methods of an object or the local variables within a function. But note that not all objects have a __dict__ attribute. For example, built-in types such as lists and dictionaries do not have this attribute.

Next, we add a Prototype class that implements the prototype design pattern. At the heart of this class, we have the clone() method, which is in charge of cloning the object using the copy.deepcopy() function.

> **Note**
>
> When we clone an object using `copy.deepcopy()`, the memory address of the clone must be different from the memory address of the original object.

Since cloning means that we allow setting values for optional attributes, notice how we use the `setattr` technique here with the `attrs` dictionary. Also, for more convenience, the `Prototype` class contains the `register()` and `unregister()` methods, which can be used to keep track of the cloned objects in a registry (a dictionary). The code of that class is as follows:

```python
class Prototype:
    def __init__(self):
        self.registry = {}

    def register(self, identifier: int, obj: object):
        self.registry[identifier] = obj

    def unregister(self, identifier: int):
        del self.registry[identifier]

    def clone(self, identifier: int, **attrs) -> object:
        found = self.registry.get(identifier)
        if not found:
            raise ValueError(
                f"Incorrect object identifier: {identifier}"
            )
        obj = copy.deepcopy(found)
        for key in attrs:
            setattr(obj, key, attrs[key])

        return obj
```

In the `main()` function, which we define next, we complete the program: we clone a first `Website` instance, `site1`, to get a second object `site2`. Basically, we instantiate the `Prototype` class and we use its `.clone()` method. Then, we display the result. The code for that function is as follows:

```python
def main():
    keywords = (
        "python",
        "programming",
        "scripting",
        "data",
        "automation",
    )
```

```
site1 = Website(
    "Python",
    domain="python.org",
    description="Programming language and ecosystem",
    category="Open Source Software",
    keywords=keywords,
)

proto = Prototype()
proto.register("python-001", site1)

site2 = proto.clone(
    "python-001",
    name="Python Package Index",
    domain="pypi.org",
    description="Repository for published packages",
    category="Open Source Software",
)

for site in (site1, site2):
    print(site)
```

Finally, we call the main() function, as follows:

```
if __name__ == "__main__":
    main()
```

Here is a summary of what we do in the code (ch03/prototype.py):

1. We start by importing the copy module.

2. We define a Website class, with its initialization method (__init__()) and its string representation method (__str__()).

3. We define our Prototype class as shown earlier.

4. Then, we have the main() function, where we do the following:

    - We define a keywords list we need.

    - We create an instance of the Website class, called site1 (we use the keywords list here).

    - We create a Prototype object and we use its register() method to register site1 with its identifier (this helps us keep track of the cloned objects in a dictionary).

    - We clone the site1 object to get site2.

    - We display the result (both Website objects).

A sample output when I execute the `python ch03/prototype.py` command on my computer is as follows:

```
- Python (ID: 4369628560)
category: Open Source Software
description: Programming language and ecosystem
domain: python.org
keywords: ('python', 'programming', 'scripting', 'data', 'automation')

- Python Package Index (ID: 4369627552)
category: Open Source Software
description: Repository site for Python's published packages
domain: pypi.org
keywords: ('python', 'programming', 'scripting', 'data', 'automation')
```

Indeed, `Prototype` works as expected. We can see information about the original `Website` object and its clone.

And looking at the ID value for each `Website` object, we can see that the two addresses are different.

The singleton pattern

One of the original design patterns for OOP, the singleton pattern restricts the instantiation of a class to *one* object, which is useful when you need one object to coordinate actions for the system.

The basic idea is that only one instance of a particular class, doing a job, is created for the needs of the program. To ensure that this works, we need mechanisms that prevent the instantiation of the class more than once and also prevent cloning.

In the Python programmer community, the singleton pattern is actually considered an anti-pattern. Let's explore the pattern first, and later we will discuss the alternative approaches we are encouraged to use in Python.

Real-world examples

In a real-life scenario, we can think of the captain of a ship or a boat. On the ship, they are the ones in charge. They are responsible for important decisions, and a number of requests are directed to them because of this responsibility.

Another example is the printer spooler, in an office environment, which ensures that print jobs are coordinated through a single point, avoiding conflicts and ensuring orderly printing.

Use cases for the singleton pattern

The singleton design pattern is useful when you need to create only one object or you need some sort of object capable of maintaining a global state for your program.

Other possible use cases are the following:

- Controlling concurrent access to a shared resource—for example, the class managing the connection to a database
- A service or resource that is transversal in the sense that it can be accessed from different parts of the application or by different users and do its work—for example, the class at the core of a logging system or utility

Implementing the singleton pattern

As discussed, the singleton pattern ensures that a class has only one instance and provides a global point to access it. In this example, we'll create a URLFetcher class that fetches content from web pages. We want to ensure that only one instance of this class exists to keep track of all fetched URLs.

Imagine you have multiple fetchers in different parts of your program, but you want to keep track of all URLs that have been fetched. This is a classic case for a singleton pattern. By ensuring that all parts of your program use the same fetcher instance, you can easily keep track of all fetched URLs in one place.

Initially, we create a naive version of the URLFetcher class. This class has a fetch() method that fetches the web page content and stores the URL in a list:

```python
import urllib.request


class URLFetcher:
    def __init__(self):
        self.urls = []

    def fetch(self, url):
        req = urllib.request.Request(url)
        with urllib.request.urlopen(req) as response:
            if response.code == 200:
                page_content = response.read()
            with open("content.html", "a") as f:
                f.write(page_content + "\n")
            self.urls.append(url)
```

To check if our class is a **singleton**, we can compare two instances of the class using the `is` operator. If they are the same, then it's a singleton:

```
if __name__ == "__main__":
    print(URLFetcher() is URLFetcher())
```

If you run this code (`ch03/singleton/before_singleton.py`), you'll see that the output is the following:

```
False
```

This output shows that the class in this version does not yet respect the singleton pattern. To make it a singleton, we'll use the **metaclass** technique.

> **Additional information**
>
> A metaclass in Python is a class of a class that defines how a class behaves.

We'll create a `SingletonType` metaclass that ensures that only one instance of `URLFetcher` exists, as follows:

```
import urllib.request


class SingletonType(type):
    _instances = {}
    def __call__(cls, *args, **kwargs):
        if cls not in cls._instances:
            obj = super(SingletonType, cls).__call__(*args, **kwargs)
            cls._instances[cls] = obj
        return cls._instances[cls]
```

Now, we modify our `URLFetcher` class to use this metaclass, as follows:

```
class URLFetcher(metaclass=SingletonType):
    def __init__(self):
        self.urls = []

    def fetch(self, url):
        req = urllib.request.Request(url)
        with urllib.request.urlopen(req) as response:
            if response.code == 200:
                page_content = response.read()
                with open("content.html", "a") as f:
```

```
                        f.write(str(page_content))
                self.urls.append(url)
```

Finally, we create a `main()` function and call it to test our singleton, with the following code:

```
def main():

    my_urls = [
            "http://python.org",
            "https://planetpython.org/",
            "https://www.djangoproject.com/",
    ]

    print(URLFetcher() is URLFetcher())

    fetcher = URLFetcher()
    for url in my_urls:
        fetcher.fetch(url)

    print(f"Done URLs: {fetcher.urls}")


if __name__ == "__main__":
    main()
```

Here is a summary of what we do in the code (`ch03/singleton/singleton.py`):

1. We start with our needed module imports (`urllib.request`).

2. We define a `SingletonType` class, with its special `__call__()` method.

3. We define `URLFetcher`, the class implementing the fetcher for the web pages, initializing it with the `urls` attribute; as discussed, we add its `fetch()` method.

4. Lastly, we add our `main()` function, and we add Python's conventional snippet used to call it.

To test the implementation, run the `python ch03/singleton/singleton.py` command. You should get the following output:

```
True
Done URLs: ['http://python.org', 'https://planetpython.org/',
'https://www.djangoproject.com/']
```

In addition, you will find that a file called `content.html` has been created, with the HTML text that comes from the different URLs added to it.

So, the program did its job as expected. This is a demonstration of how the singleton pattern may be used.

Should you use the singleton pattern?

While the singleton pattern has its merits, it may not always be the most Pythonic approach to managing global states or resources. Our implementation example worked, but if we stop a minute to analyze the code again, we notice the following:

- The techniques used for the implementation are rather advanced and not easy to explain to a beginner
- By reading the `SingletonType` class definition, it is not easy to immediately see that it provides a metaclass for a singleton if the name does not suggest it

In Python, developers often prefer a simpler alternative to singleton: using a module-level global object.

> **Note**
> Python modules act as natural namespaces that can contain variables, functions, and classes, making them ideal for organizing and sharing global resources.

By adopting the global object technique, as explained by Brandon Rhodes in what he calls the *Global Object Pattern* (`https://python-patterns.guide/python/module-globals/`), you can achieve the same result as the singleton pattern without the need for complex instantiation processes or forcing a class to only have one instance.

As an exercise, you can re-write the implementation of our example using a global object. For reference, the equivalent code, defining a global object, is provided in the `ch03/singleton/instead_of_singleton/example.py` file; for its use, check the `ch03/singleton/instead_of_singleton/use_example.py` file.

The object pool pattern

The object pool pattern is a creational design pattern that allows you to reuse existing objects instead of creating new ones when they are needed. This pattern is particularly useful when the cost, in terms of system resources, time, and so on of initializing a new object is high.

Real-world examples

Consider a car rental service. When a customer rents a car, the service doesn't manufacture a new car for them. Instead, it provides one from a pool of available cars. Once the customer returns the car, it goes back into the pool, ready to be used by the next customer.

Another example would be a public swimming pool. Rather than filling the pool with water every time someone wants to swim, the water is treated and reused for multiple swimmers. This saves both time and resources.

Use cases for the object pool pattern

The object pool pattern is especially useful in scenarios where resource initialization is costly or time-consuming. This could be in terms of CPU cycles, memory usage, or even network bandwidth. For example, in a shooting video game, you might use this pattern to manage bullet objects. Creating a new bullet every time a gun is fired could be resource-intensive. Instead, you could have a pool of bullet objects that are reused.

Implementing the object pool pattern

Let's implement a pool of reusable `car` objects, for a car rental application, to avoid creating and destroying them repeatedly.

First, we need to define a `Car` class, as follows:

```
class Car:
    def __init__(self, make: str, model: str):
        self.make = make
        self.model = model
        self.in_use = False
```

Then, we start defining a `CarPool` class with its initialization, as follows:

```
class CarPool:
    def __init__(self):
        self._available = []
        self._in_use = []
```

We need to express what happens when a client acquires a car. For that, we define a method on the class doing the following: if no car is available, we instantiate one and add it to the list of available cars in the pool; else, we return an available `car` object, while doing the following:

- Setting the `_in_use` attribute of the `car` object to `True`

- Adding the `car` object to the list of "in use" cars (stored in the `_in_use` attribute of the `pool` object)

We add the code of that method to the class as follows:

```
    def acquire_car(self) -> Car:
        if len(self._available) == 0:
            new_car = Car("BMW", "M3")
            self._available.append(new_car)
        car = self._available.pop()
        self._in_use.append(car)
```

```
        car.in_use = True
        return car
```

We then add a method that handles things when a client releases a car, as follows:

```
    def release_car(self, car: Car) -> None:
        car.in_use = False
        self._in_use.remove(car)
        self._available.append(car)
```

Finally, we add some code for testing the result of the implementation, as follows:

```
if __name__ == "__main__":
    pool = CarPool()
    car_name = "Car 1"

    print(f"Acquire {car_name}")
    car1 = pool.acquire_car()
    print(f"{car_name} in use: {car1.in_use}")

    print(f"Now release {car_name}")
    pool.release_car(car1)
    print(f"{car_name} in use: {car1.in_use}")
```

Here is a summary of what we do in the code (in file ch03/object_pool.py):

1. We define a Car class.
2. We define a CarPool class with the acquire_car() and release_car() methods, as shown earlier.
3. We add code for testing the result of the implementation, as shown earlier.

To test the program, run the following command:

```
python ch03/object_pool.py
```

You should get the following output:

```
Acquire Car 1
Car 1 in use: True
Now release Car 1
Car 1 in use: False
```

Well done! This output shows that our object pool pattern implementation works as intended.

Summary

In this chapter, we have seen *creational design patterns*, which are essential for crafting flexible, maintainable, and modular code. We kicked off the chapter by examining two variations of the factory pattern, each offering unique advantages for object creation. Next, we navigated through the builder pattern, which provides a more readable and maintainable way to construct complex objects. The prototype pattern followed, introducing a method to clone objects efficiently. Finally, we rounded out the chapter by discussing the singleton and object pool patterns, both of which are geared toward optimizing resource management and ensuring consistent state across the application.

Now, equipped with these foundational patterns for object creation, we are well prepared for the next chapter, where we will discover *structural design patterns*.

4

Structural Design Patterns

In the previous chapter, we covered creational patterns and object-oriented programming patterns that help us with object-creation procedures. The next category of pattern we want to present is *structural design patterns*. A structural design pattern proposes a way of composing objects to provide new functionality.

In this chapter, we're going to cover the following main topics:

- The adapter pattern
- The decorator pattern
- The bridge pattern
- The facade pattern
- The flyweight pattern
- The proxy pattern

At the end of this chapter, you will be equipped with the skills to structure your code efficiently and elegantly using structural design patterns.

Technical requirements

See the requirements presented in *Chapter 1*.

The adapter pattern

The **adapter** pattern is a structural design pattern that helps us make two incompatible interfaces compatible. What does that really mean? If we have an old component and we want to use it in a new system, or a new component that we want to use in an old system, the two can rarely communicate without requiring any code changes. But changing the code is not always possible, either because we don't have access to it, or because it is impractical. In such cases, we can write an extra layer that makes all the required modifications for enabling communication between the two interfaces. This layer is called an **adapter**.

In general, if you want to use an interface that expects `function_a()`, but you only have `function_b()`, you can use an adapter to convert (adapt) `function_b()` to `function_a()`.

Real-world examples

When you are traveling from most European countries to the UK or the USA, or the other way around, you need to use a plug adapter for charging your laptop. The same kind of adapter is needed for connecting some devices to your computer: the USB adapter.

In the software category, the `zope.interface` package (`https://pypi.org/project/zope.interface/`), part of the **Zope Toolkit (ZTK)**, provides tools that help define interfaces and perform interface adaptation. These tools are used in the core of several Python web framework projects (including Pyramid and Plone).

> **Note**
>
> `zope.interface` was the solution for working with interfaces in Python, proposed by the team (`https://zope.dev/`) behind the Zope application server and the ZTK before Python introduced built-in mechanisms, with **abstract base classes** (also called **ABCs**) first and protocols later.

Use cases for the adapter pattern

Usually, one of the two incompatible interfaces is either foreign or old/legacy. If the interface is foreign, it means that we have no access to the source code. If it is old, it is usually impractical to refactor it.

Using an adapter to make things work after they have been implemented is a good approach because it does not require access to the source code of the foreign interface. It is also often a pragmatic solution if we have to reuse some legacy code. That being said, be aware that it can introduce side effects that are hard to debug. So, use it with caution.

Implementing the adapter pattern – adapt a legacy class

Let's consider an example where we have a legacy payment system and a new payment gateway. The adapter pattern can make them work together without changing the existing code, as we are going to see.

The legacy payment system is implemented using a class, with a `make_payment()` method doing the core of the payment job, as follows:

```
class OldPaymentSystem:
    def __init__(self, currency):
        self.currency = currency

    def make_payment(self, amount):
        print(
```

```
                f"[OLD] Pay {amount} {self.currency}"
        )
```

The new payment system is implemented using the following class, providing an `execute_payment()` method:

```
class NewPaymentGateway:
    def __init__(self, currency):
        self.currency = currency

    def execute_payment(self, amount):
        print(
            f"Execute payment of {amount} {self.currency}"
        )
```

Now, we are going to add a class that will provide the **adaptation**. Our adapter class has an attribute system to store the object representing the payment system we need to adapt, which we call the **adaptee**. It also has a `make_payment()` method, where we call the `execute_payment()` method on the adaptee object to get the payment done. The code is as follows:

```
class PaymentAdapter:
    def __init__(self, system):
        self.system = system

    def make_payment(self, amount):
        self.system.execute_payment(amount)
```

This is how the `PaymentAdapter` class adapts the interface of `NewPaymentGateway` to match that of `OldPaymentSystem`.

Let's see the result of this adaptation by adding a `main()` function with testing code, as follows:

```
def main():
    old_system = OldPaymentSystem("euro")
    print(old_system)
    new_system = NewPaymentGateway("euro")
    print(new_system)

    adapter = PaymentAdapter(new_system)
    adapter.make_payment(100)
```

Let's recapitulate the complete code (see the `ch04/adapter/adapt_legacy.py` file) of the implementation:

1. We have some code for the legacy payment system, represented by the `OldPaymentSystem` class, providing a `make_payment()` method.

2. We introduce the new payment system, with the `NewPaymentGateway` class, providing an `execute_payment()` method.

3. We add a class for the adapter, `PaymentAdapter`, which has an attribute to store the payment system object and a `make_payment()` method; in that method, we call the `execute_payment()` method on the payment system object (via `self.system.execute_payment(amount)`).

4. We add code for testing our interface adaptation design (and call it within the usual `if __name__ == "__main__"` block).

Executing the code, using `python ch04/adapter/adapt_legacy.py`, should give the following output:

```
<__main__.OldPaymentSystem object at 0x10ee58fd0>
<__main__.NewPaymentGateway object at 0x10ee58f70>
Execute payment of 100 euro
```

You now get the idea. This adaptation technique allows us to use the new payment gateway with existing code that expects the old interface.

Implementing the adapter pattern – adapt several classes into a unified interface

Let's look at another application to illustrate adaptation: a club's activities. Our club has two main activities:

- Hire talented artists to perform in the club

- Organize performances and events to entertain its clients

At the core, we have a `Club` class that represents the club where hired artists perform some evenings. The `organize_performance()` method is the main action that the club can perform. The code is as follows:

```
class Club:
    def __init__(self, name):
        self.name = name

    def __str__(self):
        return f"the club {self.name}"

    def organize_event(self):
        return "hires an artist to perform"
```

Most of the time, our club hires a DJ to perform, but our application should make it possible to organize a diversity of performances: by a musician or music band, by a dancer, a one-man or one-woman show, and so on.

Via our research to try and reuse existing code, we find an open source contributed library that brings us two interesting classes: Musician and Dancer. In the Musician class, the main action is performed by the play() method. In the Dancer class, it is performed by the dance() method.

In our example, to indicate that these two classes are external, we place them in a separate module (in the ch04/adapter/external.py file). The code includes two classes, Musician and Dancer, as follows:

```python
class Musician:
    def __init__(self, name):
        self.name = name

    def __str__(self):
        return f"the musician {self.name}"

    def play(self):
        return "plays music"


class Dancer:
    def __init__(self, name):
        self.name = name

    def __str__(self):
        return f"the dancer {self.name}"

    def dance(self):
        return "does a dance performance"
```

The code we are writing, to use these two classes from the external library, only knows how to call the organize_performance() method (on the Club class); it has no idea about the play() or dance() methods (on the respective classes).

How can we make the code work without changing the Musician and Dancer classes?

Adapters to the rescue! We create a generic Adapter class that allows us to adapt a number of objects with different interfaces into one unified interface. The obj argument of the __init__() method is the object that we want to adapt, and adapted_methods is a dictionary containing key/value pairs matching the method the client calls and the method that should be called. The code for that class is as follows:

```python
class Adapter:
    def __init__(self, obj, adapted_methods):
        self.obj = obj
```

```
        self.__dict__.update(adapted_methods)

    def __str__(self):
        return str(self.obj)
```

When dealing with the instances of the different classes, we have two cases:

- The compatible object that belongs to the Club class needs no adaptation. We can treat it as is.

- The incompatible objects need to be adapted first, using the Adapter class.

The result is that the client code can continue using the known organize_performance() method on all objects without the need to be aware of any interface differences. Consider the following main() function code to prove that the design works as expected:

```
def main():
    objects = [
        Club("Jazz Cafe"),
        Musician("Roy Ayers"),
        Dancer("Shane Sparks"),
    ]

    for obj in objects:
        if hasattr(obj, "play") or hasattr(
            obj, "dance"
        ):
            if hasattr(obj, "play"):
                adapted_methods = dict(
                    organize_event=obj.play
                )
            elif hasattr(obj, "dance"):
                adapted_methods = dict(
                    organize_event=obj.dance
                )

            obj = Adapter(obj, adapted_methods)

        print(f"{obj} {obj.organize_event()}")
```

Let's recapitulate the complete code of our adapter pattern implementation (in the ch04/adapter/adapt_to_unified_interface.py file):

1. We import the Musician and Dancer classes from the external module.

2. We have the Club class.

3. We define the `Adapter` class.

4. We add the `main()` function, which we call within the usual `if __name__ == "__main__"` block.

Here is the output when executing the `python ch04/adapter/adapt_to_unified_interface.py` command:

```
the club Jazz Cafe hires an artist to perform
the musician Roy Ayers plays music
the dancer Shane Sparks does a dance performance
```

As you can see, we managed to make the `Musician` and `Dancer` classes compatible with the interface expected by the client code without changing the source code of these external classes.

The decorator pattern

A second interesting structural pattern to learn about is the **decorator** pattern, which allows a programmer to add responsibilities to an object dynamically, and in a transparent manner (without affecting other objects).

There is another reason why this pattern is interesting to us, as you will see in a minute.

As Python developers, we can write decorators in a **Pythonic** way (meaning using the language's features), thanks to the built-in decorator feature.

> **Note**
>
> A Python decorator is a callable (function, method, or class) that gets a `func_in` function object as input and returns another function object, `func_out`. It is a commonly used technique for extending the behavior of a function, method, or class.
>
> For more details on Python's decorator feature, see the official documentation: `https://docs.python.org/3/reference/compound_stmts.html#function`

But this feature should not be completely new to you. We have already encountered commonly used decorators in previous chapters (`@abstractmethod`, `@property`) and there are several other useful built-in decorators in Python. Now, we are going to learn how to implement and use our own decorators.

Note that there is no one-to-one relationship between the decorator pattern and Python's decorator feature. Python decorators can actually do much more than the decorator pattern. One of the things they can be used for is to implement the decorator pattern.

Real-world examples

The decorator pattern is generally used for extending the functionality of an object. In everyday life, examples of such extensions are adding a silencer to a gun, using different camera lenses, and so on.

In web frameworks such as Django, which uses decorators a lot, we have decorators that can be used for the following:

- Restricting access to views (or HTTP-request-handling functions) based on the request
- Controlling the caching behavior on specific views
- Controlling compression on a per-view basis
- Controlling caching based on specific HTTP request headers
- Registering a function as an event subscriber
- Protecting a function with a specific permission

Use cases for the decorator pattern

The decorator pattern shines when used for implementing cross-cutting concerns, such as the following:

- Data validation
- Caching
- Logging
- Monitoring
- Debugging
- Business rules
- Encryption

In general, all parts of an application that are generic and can be applied to many other parts of it are considered to be cross-cutting concerns.

Another popular example of using the decorator pattern is in **graphical user interface (GUI)** toolkits. In a GUI toolkit, we want to be able to add features such as borders, shadows, colors, and scrolling to individual components/widgets.

Implementing the decorator pattern

Python decorators are generic and very powerful. In this section, we will see how we can implement a **memoization** decorator. All recursive functions can benefit from memoization, so let's try a `number_sum()` function that returns the sum of the first *n* numbers. Note that this function is already available in the `math` module as `fsum()`, but let's pretend it is not.

First, let's look at the naive implementation (in the `ch04/decorator/number_sum_naive.py` file):

```
def number_sum(n):
    if n == 0:
        return 0
    else:
        return n + number_sum(n - 1)

if __name__ == "__main__":
    from timeit import Timer

    t = Timer(
        "number_sum(50)",
        "from __main__ import number_sum",
    )
    print("Time: ", t.timeit())
```

A sample execution of this example shows how slow this implementation is. On my computer, it takes more than *7* seconds to calculate the sum of the first 50 numbers. We get the following output when executing the `python ch04/decorator/number_sum_naive.py` command:

```
Time:   7.286800935980864
```

Let's see whether using memoization can help us improve the performance number. In the following code, we use `dict` for caching the already computed sums. We also change the parameter passed to the `number_sum()` function. We want to calculate the sum of the first 300 numbers instead of only the first 50.

Here is the new version of the code (in the `ch04/decorator/number_sum.py` file), using memoization:

```
sum_cache = {0: 0}


def number_sum(n):
    if n in sum_cache:
        return sum_cache[n]
```

```
        res = n + number_sum(n - 1)
        # Add the value to the cache
        sum_cache[n] = res
        return res


if __name__ == "__main__":
    from timeit import Timer

    t = Timer(
        "number_sum(300)",
        "from __main__ import number_sum",
    )
    print("Time: ", t.timeit())
```

Executing the memoization-based code shows that performance improves dramatically, and is acceptable even for computing large values.

A sample execution, using `python ch04/decorator/number_sum.py`, is as follows:

```
Time:   0.1288748119986849
```

But there are a few problems with this approach. First, while the performance is not an issue any longer, the code is not as clean as it is when not using memoization. And what happens if we decide to extend the code with more math functions and turn it into a module? We can think of several functions that would be useful for our module, for problems such as Pascal's triangle or the Fibonacci numbers suite algorithm.

So, if we wanted a function in the same module as `number_sum()` for the Fibonacci numbers suite, using the same memoization technique, we would add code as follows (see the version in the `ch04/decorator/number_sum_and_fibonacci.py` file):

```
fib_cache = {0: 0, 1: 1}

def fibonacci(n):
    if n in fib_cache:
        return fib_cache[n]

    res = fibonacci(n - 1) + fibonacci(n - 2)
    fib_cache[n] = res
    return res
```

Do you notice the problem? We ended up with a new dictionary called `fib_cache` that acts as our cache for the `fibonacci()` function, and a function that is more complex than it would be without using memoization. Our module is becoming unnecessarily complex.

Is it possible to write these functions while keeping them as simple as the naive versions, but achieving a performance similar to the performance of the functions that use memoization?

Fortunately, it is, and the solution is to use the decorator pattern.

First, we create a memoize() decorator as shown in the following example. Our decorator accepts the func function, which needs to be memoized, as an input. It uses dict named cache as the cached data container. The functools.wraps() function is used for convenience when creating decorators. It is not mandatory but it's a good practice to use it, since it makes sure that the documentation and the signature of the function that is decorated are preserved. The *args argument list is required in this case because the functions that we want to decorate accept input arguments (such as the n argument for our two functions):

```python
import functools

def memoize(func):
    cache = {}


    @functools.wraps(func)
    def memoizer(*args):
        if args not in cache:
            cache[args] = func(*args)
        return cache[args]

    return memoizer
```

Now we can use our memoize() decorator with the naive version of our functions. This has the benefit of readable code without performance impact. We apply a decorator using what is known as **decoration** (or a **decoration line**). A decoration uses the @name syntax, where name is the name of the decorator that we want to use. It is nothing more than syntactic sugar for simplifying the usage of decorators. We can even bypass this syntax and execute our decorator manually, but that is left as an exercise for you.

So, the memoize() decorator can be used with our recursive functions as follows:

```python
@memoize
def number_sum(n):
    if n == 0:
        return 0
    else:
        return n + number_sum(n - 1)


@memoize
```

```
def fibonacci(n):
    if n in (0, 1):
        return n
    else:
        return fibonacci(n - 1) + fibonacci(n - 2)
```

In the last part of the code, via the `main()` function, we show how to use the decorated functions and measure their performance. The `to_execute` variable is used to hold a list of tuples containing the reference to each function and the corresponding `timeit.Timer()` call (to execute it while measuring the time spent), thus avoiding code repetition. Note how the `__name__` and `__doc__` method attributes show the proper function names and documentation values, respectively. Try removing the `@functools.wraps(func)` decoration from `memoize()` and see whether this is still the case.

Here is the last part of the code:

```
def main():
    from timeit import Timer

    to_execute = [
        (
            number_sum,
            Timer(
                "number_sum(300)",
                "from __main__ import number_sum",
            ),
        ),
        (
            fibonacci,
            Timer(
                "fibonacci(100)",
                "from __main__ import fibonacci",
            ),
        ),
    ]

    for item in to_execute:
        func = item[0]
        print(
            f'Function "{func.__name__}": {func.__doc__}'
        )
        t = item[1]
        print(f"Time: {t.timeit()}")
        print()
```

Let's recapitulate how we write the complete code of our math module (the `ch04/decorator/decorate_math.py` file):

1. After the import of Python's `functools` module that we will be using, we define the `memoize()` decorator function.

2. Then, we define the `number_sum()` function, decorated using `memoize()`.

3. Next, we define the `fibonacci()` function, decorated the same way.

4. Finally, we add the `main()` function, as shown earlier, and the usual trick to call it.

Here is a sample output when executing the `python ch04/decorator/decorate_math.py` command:

```
Function "number_sum": Returns the sum of the first n numbers
Time: 0.2148694

Function "fibonacci": Returns the suite of Fibonacci numbers
Time: 0.202763251
```

> **Note**
>
> The execution times might differ in your case. Also, regardless of the time spent, we can see that the decorator-based implementation is a win because the code is more maintainable.

Nice! We ended up with readable code and acceptable performance. Now, you might argue that this is not the decorator pattern, since we don't apply it at runtime. The truth is that a decorated function cannot be undecorated, but you can still decide at runtime whether the decorator will be executed or not. That's an interesting exercise left for you. *Hint for the exercise:* use a decorator that acts as a wrapper, which decides whether or not the real decorator is executed based on some condition.

The bridge pattern

A third structural pattern to look at is the **bridge** pattern. We can actually compare the bridge and the adapter patterns, looking at the way both work. While the adapter pattern is used *later* to make unrelated classes work together, as we saw in the implementation example we discussed earlier in the section on *The adapter pattern*, the bridge pattern is designed *up-front* to decouple an implementation from its abstraction, as we are going to see.

Real-world examples

In our modern, everyday lives, an example of the bridge pattern I can think of is from the *digital economy*: information products. Nowadays, the information product or *infoproduct* is part of the resources one can find online for training, self-improvement, or one's ideas and business development. The purpose of an information product that you find on certain marketplaces, or the website of the provider, is to deliver information on a given topic in such a way that it is easy to access and consume. The provided material can be a PDF document or ebook, an ebook series, a video, a video series, an online course, a subscription-based newsletter, or a combination of all those formats.

In the software realm, we can find two examples:

- **Device drivers**: Developers of an OS define the interface for device (such as printers) vendors to implement it

- **Payment gateways**: Different payment gateways can have different implementations, but the checkout process remains consistent

Use cases for the bridge pattern

Using the bridge pattern is a good idea when you want to share an implementation among multiple objects. Basically, instead of implementing several specialized classes, and defining all that is required within each class, you can define the following special components:

- An abstraction that applies to all the classes

- A separate interface for the different objects involved

An implementation example we are about to see will illustrate this approach.

Implementing the bridge pattern

Let's assume we are building an application where the user is going to manage and deliver content after fetching it from diverse sources, which could be the following:

- A web page (based on its URL)

- A resource accessed on an FTP server

- A file on the local filesystem

- A database server

So, here is the idea: instead of implementing several content classes, each holding the methods responsible for getting the content pieces, assembling them, and showing them inside the application, we can define an abstraction for the *Resource Content* and a separate interface for the objects that are responsible for fetching the content. Let's try it!

We begin with the interface for the implementation classes that help fetch content – that is, the ResourceContentFetcher class. This concept is called the **Implementor**. Let's use Python's protocols feature, as follows:

```
class ResourceContentFetcher(Protocol):

    def fetch(self, path: str) -> str:
        ...
```

Then, we define the class for our Resource Content abstraction, called ResourceContent. The first trick we use here is that, via an attribute (_imp) on the ResourceContent class, we maintain a reference to the object that represents the Implementor (fulfilling the ResourceContentFetcher interface). The code is as follows:

```
class ResourceContent:

    def __init__(self, imp: ResourceContentFetcher):
        self._imp = imp

    def get_content(self, path):
        return self._imp.fetch(path)
```

Now we can add an implementation class to fetch content from a web page or resource:

```
class URLFetcher:

    def fetch(self, path):
        res = ""
        req = urllib.request.Request(path)
        with urllib.request.urlopen(
            req
        ) as response:
            if response.code == 200:
                res = response.read()
        return res
```

We can also add an implementation class to fetch content from a file on the local filesystem:

```
class LocalFileFetcher:

    def fetch(self, path):
        with open(path) as f:
            res = f.read()
        return res
```

Based on that, a main function with some testing code to show content using both *content fetchers* could look like the following:

```
def main():
    url_fetcher = URLFetcher()
    rc = ResourceContent(url_fetcher)
    res = rc.get_content("http://python.org")
    print(
        f"Fetched content with {len(res)} characters"
    )

    localfs_fetcher = LocalFileFetcher()
    rc = ResourceContent(localfs_fetcher)
    pathname = os.path.abspath(__file__)
    dir_path = os.path.split(pathname)[0]
    path = os.path.join(dir_path, "file.txt")
    res = rc.get_content(path)
    print(
        f"Fetched content with {len(res)} characters"
    )
```

Let's see a summary of the complete code of our example (the ch04/bridge/bridge.py file):

1. We import the modules we need for the program (os, urllib.request, and typing. Protocol).

2. We define the ResourceContentFetcher interface, using *protocols*, for the *Implementor*.

3. We define the ResourceContent class for the interface of the abstraction.

4. We define two implementation classes:

    -   URLFetcher for fetching content from a URL

    -   LocalFileFetcher for fetching content from the local filesystem

5. Finally, we add the main() function, as shown earlier, and the usual trick to call it.

Here is a sample output when executing the python ch04/bridge/bridge.py command:

```
Fetched content with 51265 characters
Fetched content with 1327 characters
```

This is a basic illustration of how using the bridge pattern in your design, you can extract content from different sources and integrate the results in the same data manipulation system or user interface.

The facade pattern

As systems evolve, they can get very complex. It is not unusual to end up with a very large (and sometimes confusing) collection of classes and interactions. In many cases, we don't want to expose this complexity to the client. This is where our next structural pattern comes to the rescue: **facade**.

The facade design pattern helps us hide the internal complexity of our systems and expose only what is necessary to the client through a simplified interface. In essence, facade is an abstraction layer implemented over an existing complex system.

Let's take the example of the computer to illustrate things. A computer is a complex machine that depends on several parts to be fully functional. To keep things simple, the word "computer," in this case, refers to an IBM derivative that uses a von Neumann architecture. Booting a computer is a particularly complex procedure. The CPU, main memory, and hard disk need to be up and running, the boot loader must be loaded from the hard disk to the main memory, the CPU must boot the operating system kernel, and so forth. Instead of exposing all this complexity to the client, we create a facade that encapsulates the whole procedure, making sure that all steps are executed in the right order.

In terms of object design and programming, we should have several classes, but only the `Computer` class needs to be exposed to the client code. The client will only have to execute the `start()` method of the `Computer` class, for example, and all the other complex parts are taken care of by the facade `Computer` class.

Real-world examples

The facade pattern is quite common in life. When you call a bank or a company, you are usually first connected to the customer service department. The customer service employee acts as a facade between you and the actual department (billing, technical support, general assistance, and so on), where an employee will help you with your specific problem.

As another example, a key used to turn on a car or motorcycle can also be considered a facade. It is a simple way of activating a system that is very complex internally. And, of course, the same is true for other complex electronic devices that we can activate with a single button, such as computers.

In software, the `django-oscar-datacash` module is a Django third-party module that integrates with the **DataCash** payment gateway. The module has a gateway class that provides fine-grained access to the various DataCash APIs. On top of that, it also offers a facade class that provides a less granular API (for those who don't want to mess with the details), and the ability to save transactions for auditing purposes.

The `Requests` library is another great example of the facade pattern. It simplifies sending HTTP requests and handling responses, abstracting the complexities of the HTTP protocol. Developers can easily make HTTP requests without dealing with the intricacies of sockets or the underlying HTTP methods.

Use cases for the facade pattern

The most usual reason to use the facade pattern is to provide a single, simple entry point to a complex system. By introducing facade, the client code can use a system by simply calling a single method/ function. At the same time, the internal system does not lose any functionality, it just encapsulates it.

Not exposing the internal functionality of a system to the client code gives us an extra benefit: we can introduce changes to the system, but the client code remains unaware of and unaffected by the changes. No modifications are required to the client code.

Facade is also useful if you have more than one layer in your system. You can introduce one facade entry point per layer and let all layers communicate with each other through their facades. That promotes **loose coupling** and keeps the layers as independent as possible.

Implementing the facade pattern

Assume that we want to create an operating system using a multi-server approach, similar to how it is done in MINIX 3 or GNU Hurd. A multi-server operating system has a minimal kernel, called the **microkernel**, which runs in privileged mode. All the other services of the system are following a server architecture (driver server, process server, file server, and so forth). Each server belongs to a different memory address space and runs on top of the microkernel in user mode. The pros of this approach are that the operating system can become more fault-tolerant, reliable, and secure. For example, since all drivers are running in user mode on a driver server, a bug in a driver cannot crash the whole system, nor can it affect the other servers. The cons of this approach are the performance overhead and the complexity of system programming, because the communication between a server and the microkernel, as well as between the independent servers, happens using message passing. Message passing is more complex than the shared memory model used in monolithic kernels such as Linux.

We begin with a `Server` interface. Also, an Enum parameter describes the different possible states of a server. We use the ABC technique to forbid direct instantiation of the `Server` interface and make the fundamental `boot()` and `kill()` methods mandatory, assuming that different actions are needed to be taken for booting, killing, and restarting each server. Here is the code for these elements, the first important bits to support our implementation:

```
State = Enum(
    "State",
    "NEW RUNNING SLEEPING RESTART ZOMBIE",
)
# ...
class Server(ABC):
    @abstractmethod
    def __init__(self):
        pass
```

```
    def __str__(self):
        return self.name

    @abstractmethod
    def boot(self):
        pass

    @abstractmethod
    def kill(self, restart=True):
        pass
```

A modular operating system can have a great number of interesting servers: a file server, a process server, an authentication server, a network server, a graphical/window server, and so forth. The following example includes two stub servers: `FileServer` and `ProcessServer`. Apart from the `boot()` and `kill()` methods all servers have, `FileServer` has a `create_file()` method for creating files, and `ProcessServer` has a `create_process()` method for creating processes.

The `FileServer` class is as follows:

```
class FileServer(Server):
    def __init__(self):
        self.name = "FileServer"
        self.state = State.NEW

    def boot(self):
        print(f"booting the {self}")
        self.state = State.RUNNING

    def kill(self, restart=True):
        print(f"Killing {self}")
        self.state = (
            State.RESTART if restart else State.ZOMBIE
        )

    def create_file(self, user, name, perms):
        msg = (
            f"trying to create file '{name}' "
            f"for user '{user}' "
            f"with permissions {perms}"
        )
        print(msg)
```

The `ProcessServer` class is as follows:

```
class ProcessServer(Server):
    def __init__(self):
        self.name = "ProcessServer"
        self.state = State.NEW

    def boot(self):
        print(f"booting the {self}")
        self.state = State.RUNNING

    def kill(self, restart=True):
        print(f"Killing {self}")
        self.state = (
            State.RESTART if restart else State.ZOMBIE
        )

    def create_process(self, user, name):
        msg = (
            f"trying to create process '{name}' "
            f"for user '{user}'"
        )
        print(msg)
```

The `OperatingSystem` class is a facade. In its `__init__()`, all the necessary server instances are created. The `start()` method, used by the client code, is the entry point to the system. More wrapper methods can be added, if necessary, as access points to the services of the servers, such as the wrappers, `create_file()` and `create_process()`. From the client's point of view, all those services are provided by the `OperatingSystem` class. The client should not be confused by unnecessary details such as the existence of servers and the responsibility of each server.

The code for the `OperatingSystem` class is as follows:

```
class OperatingSystem:
    """The Facade"""

    def __init__(self):
        self.fs = FileServer()
        self.ps = ProcessServer()

    def start(self):
        [i.boot() for i in (self.fs, self.ps)]

    def create_file(self, user, name, perms):
```

```
        return self.fs.create_file(user, name, perms)

    def create_process(self, user, name):
        return self.ps.create_process(user, name)
```

As you are going to see in a minute, when we present a summary of the example, there are many dummy classes and servers. They are there to give you an idea about the required abstractions (User, Process, File, and so forth) and servers (WindowServer, NetworkServer, and so forth) for making the system functional.

Finally, we add our main code for testing the design, as follows:

```
def main():
    os = OperatingSystem()
    os.start()
    os.create_file("foo", "hello.txt", "-rw-r-r")
    os.create_process("bar", "ls /tmp")
```

We are going to recapitulate the details of our implementation example; the full code is in the ch04/ facade.py file:

1. We start with the imports we need.

2. We define the State constant using Enum, as shown earlier.

3. We then add the User, Process, and File classes, which do nothing in this minimal but functional example.

4. We define the abstract Server class, as shown earlier.

5. We then define the FileServer class and the ProcessServer class, which are both subclasses of Server.

6. We add two other dummy classes, WindowServer and NetworkServer.

7. Then we define our facade class, OperatingSystem, as shown earlier.

8. Finally, we add the main part of the code, where we use the facade we have defined.

As you can see, executing the python ch04/facade.py command shows the messages produced by our two stub servers:

```
booting the FileServer
booting the ProcessServer
trying to create file 'hello.txt' for user 'foo' with permissions -rw-
r-r
trying to create process 'ls /tmp' for user 'bar'
```

The facade `OperatingSystem` class does a good job. The client code can create files and processes without needing to know internal details about the operating system, such as the existence of multiple servers. To be precise, the client code can call the methods for creating files and processes, but they are currently dummy. As an interesting exercise, you can implement one of the two methods, or even both.

The flyweight pattern

Whenever we create a new object, extra memory needs to be allocated. Although virtual memory provides us, theoretically, with unlimited memory, the reality is different. If all the physical memory of a system gets exhausted, it will start swapping pages with the secondary storage, usually a **hard disk drive** (**HDD**), which, in most cases, is unacceptable due to the performance differences between the main memory and HDD. **Solid-state drives** (**SSDs**) generally have better performance than HDDs, but not everybody is expected to use SSDs. So, SSDs are not going to totally replace HDDs anytime soon.

Apart from memory usage, performance is also a consideration. Graphics software, including computer games, should be able to render 3-D information (for example, a forest with thousands of trees, a village full of soldiers, or an urban area with a lot of cars) extremely quickly. If each object in a 3-D terrain is created individually and no data sharing is used, the performance will be prohibitive.

As software engineers, we should solve software problems by writing better software, instead of forcing the customer to buy extra or better hardware. The **flyweight** design pattern is a technique used to minimize memory usage and improve performance by introducing data sharing between similar objects. A flyweight is a shared object that contains state-independent, immutable (also known as **intrinsic**) data. The state-dependent, mutable (also known as **extrinsic**) data should not be part of flyweight because this is information that cannot be shared, since it differs per object. If flyweight needs extrinsic data, it should be provided explicitly by the client code.

An example might help to clarify how the flyweight pattern can be used practically. Let's assume that we are creating a performance-critical game – for example, a **first-person shooter** (**FPS**). In FPS games, the players (soldiers) share some states, such as representation and behavior. In *Counter-Strike*, for instance, all soldiers on the same team (counter-terrorists versus terrorists) look the same (representation). In the same game, all soldiers (on both teams) have some common actions, such as jump, duck, and so forth (behavior). This means that we can create a flyweight that will contain all of the common data. Of course, the soldiers also have a lot of data that is different per soldier and will not be a part of the flyweight, such as weapons, health, location, and so on.

Real-world examples

Flyweight is an optimization design pattern; therefore, it is not easy to find a good non-computing example of it. We can think of flyweight as caching in real life. For example, many bookstores have dedicated shelves with the newest and most popular publications. This is a cache. First, you can take a look at the dedicated shelves for the book you are looking for, and if you cannot find it, you can ask the bookseller to assist you.

The Exaile music player uses flyweight to reuse objects (in this case, music tracks) that are identified by the same URL. There's no point in creating a new object if it has the same URL as an existing object, so the same object is reused to save resources.

Use cases for the flyweight pattern

Flyweight is all about improving performance and memory usage. All embedded systems (phones, tablets, games consoles, microcontrollers, and so forth) and performance-critical applications (games, 3-D graphics processing, real-time systems, and so forth) can benefit from it.

The *Gang of Four* (*GoF*) book lists the following requirements that need to be satisfied to effectively use the flyweight pattern:

- The application needs to use a large number of objects.

- There are so many objects that it's too expensive to store/render them. Once the mutable state is removed (because if it is required, it should be passed explicitly to flyweight by the client code), many groups of distinct objects can be replaced by relatively few shared objects.

- Object identity is not important for the application. We cannot rely on object identity because object sharing causes identity comparisons to fail (objects that appear different to the client code end up having the same identity).

Implementing the flyweight pattern

Let's see how we can implement an example featuring cars in an area. We will create a small car park to illustrate the idea, making sure that the whole output is readable in a single terminal page. However, no matter how large you make the car park, the memory allocation stays the same.

> **Memoization versus the flyweight pattern**
>
> Memoization is an optimization technique that uses a cache to avoid recomputing results that were already computed in an earlier execution step. Memoization does not focus on a specific programming paradigm such as **object-oriented programming** (**OOP**). In Python, memoization can be applied to both class methods and simple functions.
>
> Flyweight is an OOP-specific optimization design pattern that focuses on sharing object data.

Let's get started with the code for this example.

First, we need an Enum parameter that describes the three different types of car that are in the car park:

```
CarType = Enum(
    "CarType", "SUBCOMPACT COMPACT SUV"
)
```

Then, we will define the class at the core of our implementation: `Car`. The `pool` variable is the object pool (in other words, our cache). Notice that `pool` is a class attribute (a variable shared by all instances).

Using the `__new__()` special method, which is called before `__init__()`, we are converting the `Car` class to a metaclass that supports self-references. This means that `cls` references the `Car` class. When the client code creates an instance of `Car`, they pass the type of the car as `car_type`. The type of the car is used to check whether a car of the same type has already been created. If that's the case, the previously created object is returned; otherwise, the new car type is added to the pool and returned:

```python
class Car:
    pool = dict()

    def __new__(cls, car_type):
        obj = cls.pool.get(car_type, None)
        if not obj:
            obj = object.__new__(cls)
            cls.pool[car_type] = obj
            obj.car_type = car_type
        return obj
```

The `render()` method is what will be used to render a car on the screen. Notice how all the mutable information not known by flyweight needs to be explicitly passed by the client code. In this case, random `color` and the coordinates of a location (of form x, y) are used for each car.

Also, note that to make `render()` more useful, it is necessary to ensure that no cars are rendered on top of each other. Consider this as an exercise. If you want to make rendering more fun, you can use a graphics toolkit such as Tkinter, Pygame, or Kivy.

The `render()` method is defined as follows:

```python
    def render(self, color, x, y):
        type = self.car_type
        msg = f"render a {color} {type.name} car at ({x}, {y})"
        print(msg)
```

The `main()` function shows how we can use the flyweight pattern. The color of a car is a random value from a predefined list of colors. The coordinates use random values between 1 and 100. Although 18 cars are rendered, memory is allocated only for 3. The last line of the output proves that when using flyweight, we cannot rely on object identity. The `id()` function returns the memory address of an object. This is not the default behavior in Python because, by default, `id()` returns a unique ID (actually the memory address of an object as an integer) for each object. In our case, even if two objects appear to be different, they actually have the same identity if they belong to the same **flyweight family** (in this case, the family is defined by `car_type`). Of course, different identity comparisons can still be used for objects of different families, but that is possible only if the client knows the implementation details.

Our example `main()` function's code is as follows:

```python
def main():
    rnd = random.Random()
    colors = [
        "white",
        "black",
        "silver",
        "gray",
        "red",
        "blue",
        "brown",
        "beige",
        "yellow",
        "green",
    ]
    min_point, max_point = 0, 100
    car_counter = 0

    for _ in range(10):
        c1 = Car(CarType.SUBCOMPACT)
        c1.render(
            random.choice(colors),
            rnd.randint(min_point, max_point),
            rnd.randint(min_point, max_point),
        )
        car_counter += 1

    for _ in range(3):
        c2 = Car(CarType.COMPACT)
        c2.render(
            random.choice(colors),
            rnd.randint(min_point, max_point),
            rnd.randint(min_point, max_point),
        )
        car_counter += 1

    for _ in range(5):
        c3 = Car(CarType.SUV)
        c3.render(
            random.choice(colors),
            rnd.randint(min_point, max_point),
            rnd.randint(min_point, max_point),
        )
```

```
        car_counter += 1

    print(f"cars rendered: {car_counter}")
    print(
        f"cars actually created: {len(Car.pool)}"
    )


    c4 = Car(CarType.SUBCOMPACT)
    c5 = Car(CarType.SUBCOMPACT)
    c6 = Car(CarType.SUV)
    print(
        f"{id(c4)} == {id(c5)}? {id(c4) == id(c5)}"
    )
    print(
        f"{id(c5)} == {id(c6)}? {id(c5) == id(c6)}"
    )
```

Here is the recapitulation of the full code listing (the ch04/flyweight.py file) to show you how the flyweight pattern is implemented and used:

1. We need a couple of imports: random and Enum (from the enum module).

2. We define Enum for the types of cars.

3. Then we have the Car class, with its pool attribute and the __new__() and render() methods.

4. In the first part of the main function, we define some variables and render a set of subcompact cars.

5. The second part of the main function.

6. The third part of the main function.

7. Finally, we add the fourth part of the main function.

The execution of the python ch04/flyweight.py command shows the type, random color, and coordinates of the rendered objects, as well as the identity comparison results between flyweight objects of the same/different families:

```
render a gray SUBCOMPACT car at (25, 79)
render a black SUBCOMPACT car at (31, 99)
render a brown SUBCOMPACT car at (16, 74)
render a green SUBCOMPACT car at (10, 1)
render a gray SUBCOMPACT car at (55, 38)
render a red SUBCOMPACT car at (30, 45)
render a brown SUBCOMPACT car at (17, 78)
render a gray SUBCOMPACT car at (14, 21)
render a gray SUBCOMPACT car at (7, 28)
render a gray SUBCOMPACT car at (22, 50)
```

```
render a brown COMPACT car at (75, 26)
render a red COMPACT car at (22, 61)
render a white COMPACT car at (67, 87)
render a beige SUV car at (23, 93)
render a white SUV car at (37, 100)
render a red SUV car at (33, 98)
render a black SUV car at (77, 22)
render a green SUV car at (16, 51)
cars rendered: 18
cars actually created: 3
4493672400 == 4493672400? True
4493672400 == 4493457488? False
```

Do not expect to see the same output since the colors and coordinates are random, and the object identities depend on the memory map.

The proxy pattern

The **proxy** design pattern gets its name from the *proxy* (also known as **surrogate**) object used to perform an important action before accessing the actual object. There are four well-known types of proxy. They are as follows:

1. A **virtual proxy**, which uses **lazy initialization** to defer the creation of a computationally expensive object until the moment it is actually needed.

2. A **protection/protective proxy**, which controls access to a sensitive object.

3. A **remote proxy**, which acts as the local representation of an object that really exists in a different address space (for example, a network server).

4. A **smart (reference) proxy**, which performs extra actions when an object is accessed. Examples of such actions are reference counting and thread-safety checks.

Real-world examples

Chip cards offer a good example of how a protective proxy is used in real life. The debit/credit card contains a chip that first needs to be read by the ATM or card reader. After the chip is verified, a password (PIN) is required to complete the transaction. This means that you cannot make any transactions without physically presenting the card and knowing the PIN.

A bank check that is used instead of cash to make purchases and deals is an example of a remote proxy. The check gives access to a bank account.

In software, the `weakref` module of Python contains a `proxy()` method that accepts an input object and returns a smart proxy to it. Weak references are the recommended way to add reference-counting support to an object.

Use cases for the proxy pattern

Since there are at least four common proxy types, the proxy design pattern has many use cases.

This pattern is used when creating a distributed system using either a private network or the cloud. In a distributed system, some objects exist in the local memory and some objects exist in the memory of remote computers. If we don't want the client code to be aware of such differences, we can create a remote proxy that hides/encapsulates them, making the distributed nature of the application transparent.

The proxy pattern is also handy when our application is suffering from performance issues due to the early creation of expensive objects. Introducing lazy initialization using a virtual proxy to create the objects only when they are required can give us significant performance improvements.

As a third case, this pattern is used to check whether a user has sufficient privileges to access a piece of information. If our application handles sensitive information (for example, medical data), we want to ensure that the user trying to access/modify it can do so. A protection/protective proxy can handle all security-related actions.

This pattern is used when our application (or library, toolkit, framework, and so forth) uses multiple threads and we want to move the burden of thread safety from the client code to the application. In this case, we can create a smart proxy to hide the thread-safety complexities from the client.

An **object-relational mapping** (**ORM**) API is also an example of how to use a remote proxy. Many popular web frameworks (Django, Flask, FastAPI...) use an ORM to provide OOP-like access to a relational database. An ORM acts as a proxy to a relational database that can be located anywhere, either at a local or remote server.

Implementing the proxy pattern – a virtual proxy

There are many ways to create a virtual proxy in Python, but I always like focusing on the idiomatic/Pythonic implementations. The code shown here is based on the great answer by Cyclone, a user of the `stackoverflow.com` site, to the question about "Python memoising/deferred lookup property decorator."

> **Note**
> In this section, the terms *property*, *variable*, and *attribute* are used interchangeably.

First, we create a `LazyProperty` class that can be used as a decorator. When it decorates a property, `LazyProperty` loads the property lazily (on the first use) instead of instantly. The `__init__()` method creates two variables that are used as aliases to the method that initializes a property: `method` is an alias to the actual method, and `method_name` is an alias to the method's name. To get a better understanding of how the two aliases are used, print their value to the output (uncomment the two commented lines in that part of the code):

```python
class LazyProperty:
    def __init__(self, method):
        self.method = method
        self.method_name = method.__name__
        # print(f"function overriden: {self.method}")
        # print(f"function's name: {self.method_name}")
```

The `LazyProperty` class is actually a descriptor. Descriptors are the recommended mechanisms to use in Python to override the default behavior of its attribute access methods: `__get__()`, `__set__()`, and `__delete__()`. The `LazyProperty` class overrides only `__set__()` because that is the only access method it needs to override. In other words, we don't have to override all access methods. The `__get__()` method accesses the value of the property the underlying method wants to assign, and uses `setattr()` to do the assignment manually. What `__get()__` actually does is very neat: it replaces the method with the value! This means that not only is the property lazily loaded, but it can also be set only once. We will see what this means in a moment.

```python
    def __get__(self, obj, cls):
        if not obj:
            return None
        value = self.method(obj)
        # print(f'value {value}')
        setattr(obj, self.method_name, value)
        return value
```

Again, uncomment the commented line in that part of the code to get some extra information.

Then, the `Test` class shows how we can use the `LazyProperty` class. There are three attributes: x, y, and _resource. We want the _resource variable to be loaded lazily; thus, we initialize it to None as shown in the following code:

```python
class Test:
    def __init__(self):
        self.x = "foo"
        self.y = "bar"
        self._resource = None
```

The `resource()` method is decorated with the `LazyProperty` class. For demonstration purposes, the `LazyProperty` class initializes the `_resource` attribute as a tuple, as shown in the following code. Normally, this would be a slow/expensive initialization (database, graphics, and so on):

```
@LazyProperty
def resource(self):
    print("initializing self._resource...")
    print(f"... which is: {self._resource}")
    self._resource = tuple(range(5))
    return self._resource
```

The `main()` function, as follows, shows how lazy initialization behaves:

```
def main():
    t = Test()
    print(t.x)
    print(t.y)
    # do more work...
    print(t.resource)
    print(t.resource)
```

Notice how overriding the `__get__()` access method makes it possible to treat the `resource()` method as a simple attribute (we can use `t.resource` instead of `t.resource()`).

Let's recapitulate the example code (in `ch04/proxy/proxy_lazy.py`):

1. We define the `LazyProperty` class.

2. We define the `Test` class with a `resource()` method that we decorate using `LazyProperty`.

3. We add the main function for testing our design example.

If you can execute the example in its original version (where the added lines for better understanding are kept commented), using the `python ch04/proxy/proxy_lazy.py` command, you will get the following output:

```
foo
bar
initializing self._resource...
... which is: None
(0, 1, 2, 3, 4)
(0, 1, 2, 3, 4)
```

Based on this output, we can see the following:

*   The `_resource` variable is indeed initialized not by the time the `t` instance is created, but the first time that we use `t.resource`.

- The second time t.resource is used, the variable is not initialized again. That's why the initialization string initializing self._resource is shown only once.

> **Additional information**
>
> There are two basic kinds of lazy initialization in OOP. They are as follows:
>
> - **At the instance level**: This means that an object's property is initialized lazily, but the property has an object scope. Each instance (object) of the same class has its own (different) copy of the property.
>
> - **At the class or module level**: In this case, we do not want a different copy per instance, but all the instances share the same property, which is lazily initialized. This case is not covered in this chapter. If you find it interesting, consider it as an exercise.

Since there are so many possible cases for using the proxy pattern, let's see another example.

Implementing the proxy pattern – a protection proxy

As a second example, let's implement a simple protection proxy to view and add users. The service provides two options:

- **Viewing the list of users**: This operation does not require special privileges
- **Adding a new user**: This operation requires the client to provide a special secret message

The SensitiveInfo class contains the information that we want to protect. The users variable is the list of existing users. The read() method prints the list of the users. The add() method adds a new user to the list. The code for that class is as follows:

```
class SensitiveInfo:
    def __init__(self):
        self.users = ["nick", "tom", "ben", "mike"]

    def read(self):
        nb = len(self.users)
        print(f"There are {nb} users: {' '.join(self.users)}")

    def add(self, user):
        self.users.append(user)
        print(f"Added user {user}")
```

The Info class is a protection proxy of SensitiveInfo. The secret variable is the message required to be known/provided by the client code to add a new user.

Note that this is just an example. In reality, you should never do the following:

- Store passwords in the source code

- Store passwords in clear-text form

- Use a weak (for example, MD5) or custom form of encryption

In the `Info` class, as we can see next, the `read()` method is a wrapper to `SensitiveInfo`. `read()` and the `add()` method ensures that a new user can be added only if the client code knows the secret message:

```python
class Info:
    def __init__(self):
        self.protected = SensitiveInfo()
        self.secret = "0xdeadbeef"

    def read(self):
        self.protected.read()

    def add(self, user):
        sec = input("what is the secret? ")
        if sec == self.secret:
            self.protected.add(user)
        else:
            print("That's wrong!")
```

The `main()` function shows how the proxy pattern can be used by the client code. The client code creates an instance of the `Info` class and uses the displayed menu to read the list, add a new user, or exit the application. Let's consider the following code:

```python
def main():
    info = Info()
    while True:
        print("1. read list |==| 2. add user |==| 3. quit")
        key = input("choose option: ")
        if key == "1":
            info.read()
        elif key == "2":
            name = input("choose username: ")
            info.add(name)
        elif key == "3":
            exit()
        else:
            print(f"unknown option: {key}")
```

Let's recapitulate the full code (`ch04/proxy/proxy_protection.py`):

1. First, we define the `SensitiveInfo` class.

2. Then, we have the code for the `Info` class.

3. Finally, we add the main function with our testing code.

We can see in the following a sample output of the program when executing the `python ch04/proxy/proxy_protection.py` command:

```
1. read list |==| 2. add user |==| 3. quit
choose option: 1
There are 4 users: nick tom ben mike
1. read list |==| 2. add user |==| 3. quit
choose option: 2
choose username: tom
what is the secret? 0xdeadbeef
Added user tom
1. read list |==| 2. add user |==| 3. quit
choose option: 3
```

Have you already spotted flaws or missing features that can be addressed to improve our protection proxy example? Here are a few suggestions:

- This example has a very big security flaw. Nothing prevents the client code from bypassing the security of the application by creating an instance of `SensitiveInfo` directly. Improve the example to prevent this situation. One way is to use the `abc` module to forbid direct instantiation of `SensitiveInfo`. What other code changes are required in this case?

- A basic security rule is that we should never store clear-text passwords. Storing a password safely is not very hard as long as we know which libraries to use. If you have an interest in security, try to implement a secure way to store the secret message externally (for example, in a file or database).

- The application only supports adding new users, but what about removing an existing user? Add a `remove()` method.

Implementing the proxy pattern – a remote proxy

Imagine we are building a file management system where clients can perform operations on files stored on a remote server. The operations might include reading a file, writing to a file, and deleting a file. The remote proxy hides the complexity of network requests from the client.

We start by creating an interface that defines the operations that can be performed on the remote server, `RemoteServiceInterface`, and the class that implements it to provide the actual service, `RemoteService`.

The interface is defined as follows:

```python
from abc import ABC, abstractmethod


class RemoteServiceInterface(ABC):
    @abstractmethod
    def read_file(self, file_name):
        pass

    @abstractmethod
    def write_file(self, file_name, contents):
        pass

    @abstractmethod
    def delete_file(self, file_name):
        pass
```

The RemoteService class is defined as follows (the methods just return a string, for the sake of simplicity, but normally, you would have specific code for the file handling on the remote service):

```python
class RemoteService(RemoteServiceInterface):
    def read_file(self, file_name):
        # Implementation for reading a file from the server
        return "Reading file from remote server"

    def write_file(self, file_name, contents):
        # Implementation for writing to a file on the server
        return "Writing to file on remote server"

    def delete_file(self, file_name):
        # Implementation for deleting a file from the server
        return "Deleting file from remote server"
```

Then, we define ProxyService for the proxy. It implements the RemoteServiceInterface interface and acts as a surrogate for RemoteService, which handles communication with the latter:

```python
class ProxyService(RemoteServiceInterface):
    def __init__(self):
        self.remote_service = RemoteService()

    def read_file(self, file_name):
        print("Proxy: Forwarding read request to RemoteService")
        return self.remote_service.read_file(file_name)
```

```
    def write_file(self, file_name, contents):
        print("Proxy: Forwarding write request to RemoteService")
        return self.remote_service.write_file(file_name, contents)

    def delete_file(self, file_name):
        print("Proxy: Forwarding delete request to RemoteService")
        return self.remote_service.delete_file(file_name)
```

Clients interact with the `ProxyService` component as if it were the `RemoteService` one, unaware of the remote nature of the actual service. The proxy handles the communication with the remote service, potentially adding logging, access control, or caching. To test things, we can add the following code, based on creating an instance of `ProxyService`:

```
if __name__ == "__main__":
    proxy = ProxyService()
    print(proxy.read_file("example.txt"))
```

Let's recapitulate the implementation (the full code is in `ch04/proxy/proxy_remote.py`):

1. We start by defining the interface, `RemoteServiceInterface`, and a class that implements it, `RemoteService`.

2. Then, we define the `ProxyService` class, which also implements the `RemoteService Interface` interface.

3. Finally, we add some code for testing the proxy object.

Let's see the result of the example by running `python ch04/proxy/proxy_remote.py`:

```
Proxy: Forwarding read request to RemoteService
Reading file from remote server
```

It worked. This lightweight example was effective in showing how to implement the remote proxy use case.

Implementing the proxy pattern – a smart proxy

Let's consider a scenario where you have a shared resource in your application, such as a database connection. Every time an object accesses this resource, you want to keep track of how many references to the resource exist. Once there are no more references, the resource can be safely released or closed. A smart proxy will help manage the reference counting for this database connection, ensuring it's only closed once all references to it are released.

As in the previous example, we will need an interface, `DBConnectionInterface`, defining operations for accessing the database, and a class that represents the actual database connection, `DBConnection`.

For the interface, let's use `Protocol` (to change from the ABC way):

```python
from typing import Protocol


class DBConnectionInterface(Protocol):
    def exec_query(self, query):
        ...
```

The class for the database connection is as follows:

```python
class DBConnection:
    def __init__(self):
        print("DB connection created")

    def exec_query(self, query):
        return f"Executing query: {query}"

    def close(self):
        print("DB connection closed")
```

Then, we define the `SmartProxy` class; it also implements the `DBConnectionInterface` interface (see the `exec_query()` method). We use this class to manage reference counting and access to the `DBConnection` object. It ensures that the `DBConnection` object is created on demand when the first query is executed and is only closed when there are no more references to it. The code is as follows:

```python
class SmartProxy:
    def __init__(self):
        self.cnx = None
        self.ref_count = 0

    def access_resource(self):
        if self.cnx is None:
            self.cnx = DBConnection()
        self.ref_count += 1
        print(f"DB connection now has {self.ref_count} references.")

    def exec_query(self, query):
        if self.cnx is None:
            # Ensure the connection is created
            # if not already
            self.access_resource()
```

```
        result = self.cnx.exec_query(query)
        print(result)

        # Decrement reference count after
        # executing query
        self.release_resource()

        return result

    def release_resource(self):
        if self.ref_count > 0:
            self.ref_count -= 1
            print("Reference released...")
            print(f"{self.ref_count} remaining refs.")

        if self.ref_count == 0 and self.cnx is not None:
            self.cnx.close()
            self.cnx = None
```

Now, we can add some code to test the implementation:

```
if __name__ == "__main__":
    proxy = SmartProxy()
    proxy.exec_query("SELECT * FROM users")
    proxy.exec_query("UPDATE users SET name = 'John Doe' WHERE id =
1")
```

Let's recapitulate the implementation (the full code is in ch04/proxy/proxy_smart.py):

1. We start by defining the interface, DBConnectionInterface, and a class that implements it and represents the database connection, DBConnection.

2. Then, we define the SmartProxy class, which also implements DBConnectionInterface.

3. Finally, we add some code for testing the proxy object.

Let's see the result of the example by running python ch04/proxy/proxy_smart.py:

```
DB connection created
DB connection now has 1 references.
Executing query: SELECT * FROM users
Reference released...
0 remaining refs.
DB connection closed
DB connection created
```

```
DB connection now has 1 references.
Executing query: UPDATE users SET name = 'John Doe' WHERE id = 1
Reference released...
0 remaining refs.
DB connection closed
```

This was another demonstration of the proxy pattern. Here, it helped us implement an improved solution for scenarios where database connections are shared across different parts of an application and need to be managed carefully to avoid exhausting database resources or leaking connections.

Summary

Structural patterns are invaluable for creating clean, maintainable, and scalable code. They provide solutions for many of the challenges you'll face in daily coding.

First, the adapter pattern serves as a flexible solution for harmonizing mismatched interfaces. We can use this pattern to bridge the gap between outdated legacy systems and modern interfaces, thus promoting more cohesive and manageable software systems.

Then, we discussed the decorator pattern that we use as a convenient way of extending the behavior of an object without using inheritance. Python, with its built-in decorator feature, extends the decorator concept even more by allowing us to extend the behavior of any callable without using inheritance or composition. The decorator pattern is a great solution for implementing cross-cutting concerns because they are generic and do not fit well into the OOP paradigm. We mentioned several categories of cross-cutting concerns in the *Use cases for the decorator pattern* section. We saw how decorators can help us to keep our functions clean without sacrificing performance.

Sharing similarities with the adapter pattern, the bridge pattern is different from it in the sense that it is used up-front to define an abstraction and its implementation in a decoupled way so that both can vary independently. The bridge pattern is useful when writing software for problem domains such as operation systems and device drivers, GUIs, and website builders where we have multiple themes and we need to change the theme of a website based on certain properties. We discussed an example in the domain of content extraction and management, where we defined an interface for the abstraction, an interface for the implementor, and two implementations.

The facade pattern is ideal for providing a simple interface to client code that wants to use a complex system but does not need to be aware of the system's complexity. A computer is a facade, since all we need to do to use it is press a single button to turn it on. All the rest of the hardware complexity is handled transparently by the BIOS, the boot loader, and the other components of the system software. There are more real-life examples of facade, such as when we are connected to the customer service department of a bank or a company, and the keys that we use to turn a vehicle on. We covered an implementation of the interface used by a multi-server operating system.

In general, we use the flyweight pattern when an application needs to create a large number of computationally expensive objects that share many properties. The important point is to separate the immutable (shared) properties from the mutable ones. We saw how to implement a car renderer that supports three different car families. By providing the mutable color and x, y properties explicitly to the render() method, we managed to create only 3 different objects instead of 18. Although that might not seem like a big win, imagine if the cars were 2,000 instead of 18.

We ended with the proxy pattern. We discussed several use cases of the proxy pattern, including performance, security, and how to offer simple APIs to users. We saw an implementation example for each of the four types of proxy you generally need: virtual proxy, protective proxy, proxy to a remote service, and smart proxy.

In the next chapter, we will explore behavioral design patterns, patterns that deal with object interconnection and algorithms.

5

Behavioral Design Patterns

In the previous chapter, we covered structural patterns and **object-oriented programming** (OOP) patterns that help us create clean, maintainable, and scalable code. The next category of design patterns is **behavioral design patterns**. Behavioral patterns deal with object interconnection and algorithms.

In this chapter, we're going to cover the following main topics:

- The Chain of Responsibility pattern
- The Command pattern
- The Observer pattern
- The State pattern
- The Interpreter pattern
- The Strategy pattern
- The Memento pattern
- The Iterator pattern
- The Template pattern
- Other behavioral design patterns

At the end of this chapter, you will know how to improve your software project designs using behavioral patterns.

Technical requirements

See the requirements presented in *Chapter 1*. The additional technical requirements for the code discussed in this chapter are the following:

- For the State pattern section, install the `state_machine` module, using the command: `python -m pip install state_machine`.

- For the Interpreter pattern section, install the `pyparsing` module, using the command: `python -m pip install pyparsing`.

- For the Template pattern section, install the `cowpy` module, using the command: `python -m pip install cowpy`.

The Chain of Responsibility pattern

The Chain of Responsibility pattern offers an elegant way to handle requests by passing them through a chain of handlers. Each handler in the chain has the autonomy to decide whether it can process the request or if it should delegate it further along the chain. This pattern shines when dealing with operations that involve multiple handlers but don't necessarily require all of them to be involved.

In practice, this pattern encourages us to focus on objects and the flow of a request within an application. Notably, the client code remains blissfully unaware of the entire chain of handlers. Instead, it only interacts with the first processing element in the chain. Similarly, each processing element knows only about its immediate successor, forming a one-way relationship similar to a singly linked list. This structure is purposefully designed to achieve decoupling between the sender (client) and the receivers (processing elements).

Real-world examples

ATMs and, in general, any kind of machine that accepts/returns banknotes or coins (for example, a snack-vending machine) use the Chain of Responsibility pattern. There is always a single slot for all banknotes, as shown in the following diagram, courtesy of www.sourcemaking.com:

Figure 5.1 – Chain of Responsibility example: the ATM

When a banknote is dropped, it is routed to the appropriate receptacle. When it is returned, it is taken from the appropriate receptacle. We can think of the single slot as the shared communication medium and the different receptacles as the processing elements. The result contains cash from one or more receptacles. For example, in the preceding diagram, we see what happens when we request $175 from the ATM.

In some web frameworks, filters or middleware are pieces of code that are executed before an HTTP request arrives at a target. There is a chain of filters. Each filter performs a different action (user authentication, logging, data compression, and so forth), and either forwards the request to the next filter until the chain is exhausted, or it breaks the flow if there is an error—for example, the authentication failed three consecutive times.

Use cases for the Chain of Responsibility pattern

By using the Chain of Responsibility pattern, we provide a chance for a number of different objects to satisfy a specific request. This is useful when we don't know in advance which object should satisfy a given request. An example of this is a **purchase system**. In purchase systems, there are many approval authorities. One approval authority might be able to approve orders up to a certain value, let's say $100. If the order is for more than $100, the order is sent to the next approval authority in the chain, which can approve orders up to $200, and so forth.

Another case where the Chain of Responsibility is useful is when we know that more than one object might need to process a single request. This is what happens in event-based programming. A single event, such as a left-mouse click, can be caught by more than one listener.

It is important to note that the Chain of Responsibility pattern is not very useful if all the requests can be taken care of by a single processing element unless we really don't know which element that is. The value of this pattern is the decoupling that it offers, as we have seen in the *Loose coupling* section of *Chapter 1, Foundational Design Principles*. Instead of having a many-to-many relationship between a client and all processing elements (and the same is true regarding the relationship between a processing element and all other processing elements), a client only needs to know how to communicate with the start (head) of the chain.

Implementing the Chain of Responsibility pattern

There are many ways to implement a Chain of Responsibility in Python, but my favorite implementation is the one by Vespe Savikko (`https://legacy.python.org/workshops/1997-10/proceedings/savikko.html`). Vespe's implementation uses dynamic dispatching in a Pythonic style to handle requests.

Let's implement a simple, event-based system using Vespe's implementation as a guide. The following is the UML class diagram of the system:

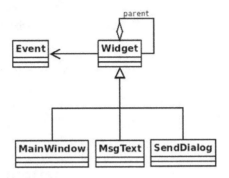

Figure 5.2 – UML class diagram of an event-based window system

The Event class describes an event. We'll keep it simple, so, in our case, an event has only name:

```
class Event:
    def __init__(self, name):
        self.name = name

    def __str__(self):
        return self.name
```

The Widget class is the core class of the application. The **parent** aggregation shown in the UML diagram indicates that each widget can have a reference to a parent object, which, by convention, we assume is a Widget instance. Note, however, that according to the rules of inheritance, an instance of any of the subclasses of Widget (for example, an instance of MsgText) is also an instance of Widget. The class has a handle() method, which uses dynamic dispatching through hasattr() and getattr() to decide who the handler of a specific request (event) is. If the widget that is asked to handle an event does not support it, there are two fallback mechanisms. If the widget has a parent, then the handle() method of the parent is executed. If the widget has no parent but a handle_default() method, handle_default() is executed. The code is as follows:

```
class Widget:
    def __init__(self, parent=None):
        self.parent = parent

    def handle(self, event):
        handler = f"handle_{event}"
        if hasattr(self, handler):
            method = getattr(self, handler)
            method(event)
        elif self.parent is not None:
```

```
            self.parent.handle(event)
        elif hasattr(self, "handle_default"):
            self.handle_default(event)
```

At this point, you might have realized why the Widget and Event classes are only associated (no aggregation or composition relationships) in the UML class diagram. The association is used to show that the Widget class knows about the Event class but does not have any strict reference to it, since an event needs to be passed only as a parameter to handle().

MainWIndow, MsgText, and SendDialog are all widgets with different behaviors. Not all these three widgets are expected to be able to handle the same events, and even if they can handle the same event, they might behave differently. MainWindow can handle only the close and default events:

```
class MainWindow(Widget):
    def handle_close(self, event):
        print(f"MainWindow: {event}")

    def handle_default(self, event):
        print(f"MainWindow Default: {event}")
```

SendDialog can handle only the paint event:

```
class SendDialog(Widget):
    def handle_paint(self, event):
        print(f"SendDialog: {event}")
```

Finally, MsgText can handle only the down event:

```
class MsgText(Widget):
    def handle_down(self, event):
        print(f"MsgText: {event}")
```

The main() function shows how we can create a few widgets and events, and how the widgets react to those events. All events are sent to all the widgets. Note the parent relationship of each widget—the sd object (an instance of SendDialog) has as its parent the mw object (an instance of MainWindow). However, not all objects need to have a parent that is an instance of MainWindow. For example, the msg object (an instance of MsgText) has the sd object as a parent:

```
def main():
    mw = MainWindow()
    sd = SendDialog(mw)
    msg = MsgText(sd)

    for e in ("down", "paint", "unhandled", "close"):
        evt = Event(e)
```

```
        print(f"Sending event -{evt}- to MainWindow")
        mw.handle(evt)
        print(f"Sending event -{evt}- to SendDialog")
        sd.handle(evt)
        print(f"Sending event -{evt}- to MsgText")
        msg.handle(evt)
```

Let's recapitulate the complete code (see file `ch05/chain.py`) of the implementation:

1. We define the `Event` class, followed by the `Widget` class.

2. We define the specialized widget classes, `MainWindow`, `SendDialog`, and `MsgText`.

3. Finally, we add the code for the `main()` function; we make sure it can be called thanks to the usual trick at the end.

Executing the `python ch05/chain.py` command gives us the following output:

```
Sending event -down- to MainWindow
MainWindow Default: down
Sending event -down- to SendDialog
MainWindow Default: down
Sending event -down- to MsgText
MsgText: down
Sending event -paint- to MainWindow
MainWindow Default: paint
Sending event -paint- to SendDialog
SendDialog: paint
Sending event -paint- to MsgText
SendDialog: paint
Sending event -unhandled- to MainWindow
MainWindow Default: unhandled
Sending event -unhandled- to SendDialog
MainWindow Default: unhandled
Sending event -unhandled- to MsgText
MainWindow Default: unhandled
Sending event -close- to MainWindow
MainWindow: close
Sending event -close- to SendDialog
MainWindow: close
Sending event -close- to MsgText
MainWindow: close
```

The Command pattern

Most applications nowadays have an undo operation. It is hard to imagine, but undo did not exist in any software for many years. Undo was introduced in 1974, but Fortran and Lisp, two programming languages that are still widely used, were created in 1957 and 1958, respectively! I wouldn't like to have been an application user during those years. Making a mistake meant that the user had no easy way to fix it.

Enough with the history. We want to know how we can implement the undo functionality in our applications. Since you have read the title of this chapter, you already know which design pattern is recommended to implement undo: the Command pattern.

The Command design pattern helps us encapsulate an operation (undo, redo, copy, paste, and so forth) as an object. What this simply means is that we create a class that contains all the logic and the methods required to implement the operation. The advantages of doing this are as follows:

- We don't have to execute a command directly. It can be executed at will.

- The object that invokes the command is decoupled from the object that knows how to perform it. The invoker does not need to know any implementation details about the command.

- If it makes sense, multiple commands can be grouped to allow the invoker to execute them in order. This is useful, for instance, when implementing a multilevel undo command.

Real-world examples

When we go to a restaurant for dinner, we give the order to the waiter. The check (usually paper) that they use to write the order is an example of a command. After writing the order, the waiter places it in the check queue that is executed by the cook. Each check is independent and can be used to execute many different commands, for example, one command for each item that will be cooked.

As you would expect, we also have several examples in the software. Here are two I can think of:

- PyQt is the Python binding of the QT toolkit. PyQt contains a `QAction` class that models an action as a command. Extra optional information is supported for every action, such as description, tooltip, and shortcut.

- Git Cola, a Git GUI written in Python, uses the command pattern to modify the model, amend a commit, apply a different election, check out, and so forth.

Use cases for the Command pattern

Many developers use the undo example as the only use case of the Command pattern. The truth is that undo is the killer feature of the Command pattern. However, the Command pattern can actually do much more:

- **GUI buttons and menu items**: The PyQt example that was already mentioned uses the Command pattern to implement actions on buttons and menu items.

- **Other operations**: Apart from undo, commands can be used to implement any operation. A few examples are *cut*, *copy*, *paste*, *redo*, and *capitalize text*.

- **Transactional behavior and logging**: Transactional behavior and logging are important to keep a persistent log of changes. They are used by operating systems to recover from system crashes, relational databases to implement transactions, filesystems to implement snapshots, and installers (wizards) to revert canceled installations.

- **Macros**: By macros, in this case, we mean a sequence of actions that can be recorded and executed on demand at any point in time. Popular editors such as Emacs and Vim support macros.

Implementing the Command pattern

Let's use the Command pattern to implement the following basic file utilities:

- Creating a file and, optionally, adding text to it
- Reading the contents of a file
- Renaming a file

We are not going to implement these utilities from scratch since Python already offers good implementations for them in the os module. What we want to do is to add an extra abstraction level on top of them so that they can be treated as commands. By doing this, we get all the advantages offered by commands.

Each command has two parts:

- **The initialization part**: It is taken care of by the __init__() method and contains all the information required by the command to be able to do something useful (the path of a file, the contents that will be written to the file, and so forth).

- **The execution part**: It is taken care of by the execute() method. We call that method when we want to run a command. This is not necessarily right after initializing it.

Let's start with the rename utility, which we implement using the `RenameFile` class. The class is initialized using the source and destination file paths. We add the `execute()` method, which does the actual renaming using `os.rename()`. To provide support for the undo operation, we add the `undo()` method, where we use `os.rename()` again to revert the name of the file to its original value. Note that we also use logging to improve the output.

The beginning of the code, the imports we need, and the `RenameFile` class, are as follows:

```python
import logging
import os

logging.basicConfig(level=logging.DEBUG)

class RenameFile:
    def __init__(self, src, dest):
        self.src = src
        self.dest = dest

    def execute(self):
        logging.info(
            f"[renaming '{self.src}' to '{self.dest}']"
        )
        os.rename(self.src, self.dest)

    def undo(self):
        logging.info(
            f"[renaming '{self.dest}' back to '{self.src}']"
        )
        os.rename(self.dest, self.src)
```

Next, we add a `CreateFile` class for the command used to create a file. The initialization method for that class accepts the familiar `path` parameter and a `txt` parameter for the content that will be written to the file. If nothing is passed as content, the default "hello world" text is written to the file. Normally, the sane default behavior is to create an empty file, but for the needs of this example, I decided to write a default string in it. Then, we add an `execute()` method, in which we use Python's `open()` function to open the file in **write mode**, and the `write()` function to write the `txt` string to it.

The undo for the operation of creating a file is to delete that file. Thus, the `undo()` method is added to the class, where we use the `os.remove()` function to do the job.

The definition for the `CreateFile` class is as follows:

```
class CreateFile:
    def __init__(self, path, txt="hello world\n"):
        self.path = path
        self.txt = txt

    def execute(self):
        logging.info(f"[creating file '{self.path}']")
        with open(
            self.path, "w", encoding="utf-8"
        ) as out_file:
            out_file.write(self.txt)

    def undo(self):
        logging.info(f"deleting file {self.path}")
        os.remove(self.path)
```

The last utility gives us the ability to read the contents of a file. The `execute()` method of the `ReadFile` class uses `open()` again, this time in read mode, and just prints the content of the file.

The `ReadFile` class is defined as follows:

```
class ReadFile:
    def __init__(self, path):
        self.path = path

    def execute(self):
        logging.info(f"[reading file '{self.path}']")
        with open(
            self.path, "r", encoding="utf-8"
        ) as in_file:
            print(in_file.read(), end="")
```

The `main()` function makes use of the utilities we have defined. The `orig_name` and `new_name` parameters are the original and new name of the file that is created and renamed. A commands list is used to add (and configure) all the commands that we want to execute at a later point. The code is as follows:

```
def main():

    orig_name, new_name = "file1", "file2"

    commands = (
        CreateFile(orig_name),
```

```
        ReadFile(orig_name),
        RenameFile(orig_name, new_name),
    )

for c in commands:
    c.execute()
```

Then, we ask the users whether they want to undo the executed commands or not. The user selects whether the commands will be undone or not. If they choose to undo them, undo() is executed for all commands in the commands list. However, since not all commands support undo, exception handling is used to catch (and log) the AttributeError exception generated when the undo() method is missing. That part of the code would look like the following:

```
answer = input("reverse the executed commands? [y/n] ")

if answer not in "yY":
    print(f"the result is {new_name}")
    exit()

for c in reversed(commands):
    try:
        c.undo()
    except AttributeError as e:
        logging.error(str(e))
```

Let's recapitulate the complete code (in the ch05/command.py file) of the implementation:

1. We import logging and os modules.
2. We do the usual logging configuration.
3. We define the RenameFile class.
4. We define the CreateFile class.
5. We define the ReadFile class.
6. We add a main() function, and call it as usual, to test our design.

Executing the python ch05/command.py command gives us the following output, if we accept to reverse the commands:

```
INFO:root:[creating file 'file1']
INFO:root:[reading file 'file1']
hello world
INFO:root:[renaming 'file1' to 'file2']
reverse the executed commands? [y/n] y
```

```
INFO:root:[renaming 'file2' back to 'file1']
ERROR:root:'ReadFile' object has no attribute 'undo'
INFO:root:deleting file file1
```

However, if we do not accept to reverse the commands, the output is as follows:

```
INFO:root:[creating file 'file1']
INFO:root:[reading file 'file1']
hello world
INFO:root:[renaming 'file1' to 'file2']
reverse the executed commands? [y/n] n
the result is file2
```

These outputs are as expected. However, note that ERROR, in the first case, is normal for this context.

The Observer pattern

The Observer pattern describes a publish-subscribe relationship between a single object, the publisher, which is also known as the subject or **observable,** and one or more objects, the subscribers, also known as **observers**. So, the subject notifies the subscribers of any state changes, typically by calling one of their methods.

The ideas behind the Observer pattern are the same as those behind the separation of concerns principle, that is, to increase decoupling between the publisher and subscribers, and to make it easy to add/remove subscribers at runtime.

Real-world examples

Dynamics in an auction are similar to the behavior of the Observer pattern. Every auction bidder has a number paddle that is raised whenever they want to place a bid. Whenever the paddle is raised by a bidder, the auctioneer acts as the subject by updating the price of the bid and broadcasting the new price to all bidders (subscribers).

In software, we can cite at least two examples:

- **Kivy**, the Python framework for developing **user interfaces (UIs)**, has a module called **Properties**, which implements the Observer pattern. Using this technique, you can specify what should happen when a property's value changes.

- The **RabbitMQ library** provides an implementation of an **Advanced Message Queuing Protocol (AMQP)** messaging broker. It is possible to construct a Python application that interacts with RabbitMQ in such a way that it subscribes to messages and publishes them to queues, which is essentially the Observer design pattern.

Use cases for the Observer pattern

We generally use the Observer pattern when we want to inform/update one or more objects (observers/subscribers) about a change that happened on a given object (subject/publisher/observable). The number of observers, as well as who those observers are, may vary and can be changed dynamically.

We can think of many cases where Observer can be useful. One such use case is news feeds. With RSS, Atom, or other related formats, you follow a feed, and every time it is updated, you receive a notification about the update.

The same concept exists in social networking applications. If you are connected to another person using a social networking service, and your connection updates something, you are notified about it.

Event-driven systems are another example where Observer is usually used. In such systems, you have listeners that listen for specific events. The listeners are triggered when an event they are listening to is created. This can be typing a specific key (on the keyboard), moving the mouse, and more. The event plays the role of the *publisher*, and the listeners play the role of the *observers*. The key point in this case is that multiple listeners (observers) can be attached to a single event (publisher).

Implementing the Observer pattern

As an example, let's implement a weather monitoring system. In such a system, you have a weather station that collects weather-related data (temperature, humidity, and atmospheric pressure). Our system needs to allow different devices and applications to receive real-time updates whenever there is a change in the weather data.

We can apply the Observer pattern using the following elements:

- **Subject (weather station):** Create a `WeatherStation` class that acts as the subject. This class will maintain a list of observers (devices or applications) interested in receiving weather updates.
- **Observers (devices and applications):** Implement various observer classes, representing devices such as smartphones, tablets, weather apps, and even a display screen in a local store. Each observer will subscribe to receive updates from the weather station.
- **Registration and notification:** The weather station provides methods for observers to register (`subscribe`) and unregister (`unsubscribe`) themselves. When there is a change in weather data (e.g., a new temperature reading), the weather station notifies all registered observers.
- **Update mechanism:** Each observer defines an `update()` method that the weather station calls when notifying about changes. For instance, a smartphone observer may update its weather app with the latest data, while a local store display may update its digital sign.

Let's get started.

First, we define the `Observer` interface, which holds an `update` method that observers must implement. Observers are expected to update themselves when the subject's state changes:

```python
class Observer:
    def update(self, temperature, humidity, pressure):
        pass
```

Next, we define the `WeatherStation` subject class. It maintains a list of observers and provides methods to add and remove observers. The `set_weather_data` method is used to simulate changes in weather data. When the weather data changes, it notifies all registered observers by calling their `update` methods. The code is as follows:

```python
class WeatherStation:
    def __init__(self):
        self.observers = []

    def add_observer(self, observer):
        self.observers.append(observer)

    def remove_observer(self, observer):
        self.observers.remove(observer)

    def set_weather_data(self, temperature, humidity, pressure):
        for observer in self.observers:
            observer.update(temperature, humidity, pressure)
```

Let's now define the `DisplayDevice` observer class. Its `update` method prints weather information when called:

```python
class DisplayDevice(Observer):
    def __init__(self, name):
        self.name = name

    def update(self, temperature, humidity, pressure):
        print(f"{self.name} Display")
        print(
            f" - Temperature: {temperature}°C, Humidity: {humidity}%,
Pressure: {pressure}hPa"
        )
```

Similarly, we define another observer class, `WeatherApp`, which prints weather information in a different format when its `update` method is called:

```python
class WeatherApp(Observer):
    def __init__(self, name):
        self.name = name

    def update(self, temperature, humidity, pressure):
        print(f"{self.name} App - Weather Update")
        print(
            f" - Temperature: {temperature}°C, Humidity: {humidity}%,
Pressure: {pressure}hPa"
        )
```

Now, in the `main()` function, we do several things:

- We create an instance of the `WeatherStation` class, which acts as the subject.
- We create instances of `DisplayDevice` and `WeatherApp`, representing different types of observers.
- We register these observers with `weather_station` using the `add_observer` method.
- We simulate changes in weather data by calling the `set_weather_data` method of `weather_station`. This triggers updates to all registered observers.

The code of the `main()` function is as follows:

```python
def main():
    # Create the WeatherStation
    weather_station = WeatherStation()

    # Create and register observers
    display1 = DisplayDevice("Living Room")
    display2 = DisplayDevice("Bedroom")
    app1 = WeatherApp("Mobile App")

    weather_station.add_observer(display1)
    weather_station.add_observer(display2)
    weather_station.add_observer(app1)

    # Simulate weather data changes
    weather_station.set_weather_data(25.5, 60, 1013.2)
    weather_station.set_weather_data(26.0, 58, 1012.8)
```

Let's recapitulate the complete code (in the `ch05/observer.py` file) of the implementation:

1. We define the Observer interface.

2. We define the `WeatherStation` subject class.

3. We define two observer classes, `DisplayDevice` and `WeatherApp`.

4. We add a `main()` function where we test our design.

Executing the `python ch05/observer.py` command gives us the following output:

```
Living Room Display
- Temperature: 25.5°C, Humidity: 60%, Pressure: 1013.2hPa
Bedroom Display
- Temperature: 25.5°C, Humidity: 60%, Pressure: 1013.2hPa
Mobile App App - Weather Update
- Temperature: 25.5°C, Humidity: 60%, Pressure: 1013.2hPa
Living Room Display
- Temperature: 26.0°C, Humidity: 58%, Pressure: 1012.8hPa
Bedroom Display
- Temperature: 26.0°C, Humidity: 58%, Pressure: 1012.8hPa
Mobile App App - Weather Update
- Temperature: 26.0°C, Humidity: 58%, Pressure: 1012.8hPa
```

As you can see, this example demonstrates the Observer pattern, where the subject notifies its observers about changes in its state. Observers are loosely coupled with the subject and can be added or removed dynamically, providing flexibility and decoupling in the system.

As an exercise, you can see that when unregistering an observer, using the `remove_observer()` method, and then simulating additional weather data changes, only the remaining registered observers receive updates. As a helper, to test this, here are 2 lines of code to add at the end of the `main()` function:

```
weather_station.remove_observer(display2)
weather_station.set_weather_data(27.2, 55, 1012.5)
```

Next, we will discuss the State pattern.

The State pattern

In the previous chapter, we covered the Observer pattern, which is useful in a program to notify other objects when the state of a given object changes. Let's continue discovering those patterns proposed by the Gang of Four.

OOP focuses on maintaining the states of objects that interact with each other. A very handy tool to model state transitions when solving many problems is known as a **finite-state machine** (commonly called a **state machine**).

What's a state machine? A state machine is an abstract machine that has two key components, that is, states and transitions. A state is the current (active) status of a system. For example, if we have a radio receiver, two possible states for it are to be tuned to FM or AM. Another possible state is for it to be switching from one FM/AM radio station to another. A transition is a switch from one state to another. A transition is initiated by a triggering event or condition. Usually, an action or set of actions is executed before or after a transition occurs. Assuming that our radio receiver is tuned to the 107 FM station, an example of a transition is for the button to be pressed by the listener to switch it to 107.5 FM.

A nice feature of state machines is that they can be represented as graphs (called **state diagrams**), where each state is a node, and each transition is an edge between two nodes.

State machines can be used to solve many kinds of problems, both non-computational and computational. Non-computational examples include vending machines, elevators, traffic lights, combination locks, parking meters, and automated gas pumps. Computational examples include game programming and other categories of computer programming, hardware design, protocol design, and programming language parsing.

Now, we have an idea of what state machines are! But how are state machines related to the State design pattern? It turns out that the State pattern is nothing more than a state machine applied to a particular software engineering problem (*Gang of Four-95*, page 342), (*Python 3 Patterns, Recipes and Idioms by Bruce Eckel & Friends*, page 151).

Real-world examples

A snack vending machine is an example of the State pattern in everyday life. Vending machines have different states and react differently depending on the amount of money that we insert. Depending on our selection and the money we insert, the machine can do the following:

- Reject our selection because the product we requested is out of stock.
- Reject our selection because the amount of money we inserted was not sufficient.
- Deliver the product and give no change because we inserted the exact amount.
- Deliver the product and return the change.

There are, for sure, more possible states, but you get the point.

Other examples of the state pattern in real life are as follows:

- Traffic lights
- Game states in a video game

In software, the state pattern is commonly used. Python and its ecosystem offer several packages/ modules one can use to implement state machines. We will see how to use one of them in the implementation section.

Use cases for the State pattern

The State pattern is applicable to many problems. All the problems that can be solved using state machines are good use cases for using the State pattern. An example we have already seen is the process model for an operating/embedded system.

Programming language compiler implementation is another good example. Lexical and syntactic analysis can use states to build abstract syntax trees.

Event-driven systems are yet another example. In an event-driven system, the transition from one state to another triggers an event/message. Many computer games use this technique. For example, a monster might move from the guard state to the attack state when the main hero approaches it.

To quote Thomas Jaeger, in his article, *The State Design Pattern vs. State Machine* (`https://thomasjaeger.wordpress.com/2012/12/13/the-state-design-pattern-vs-state-machine-2/`):

The state design pattern allows for full encapsulation of an unlimited number of states on a context for easy maintenance and flexibility.

Implementing the State pattern

Let's write code that demonstrates how to create a state machine based on the state diagram shown earlier in this chapter. Our state machine should cover the different states of a process and the transitions between them.

The State design pattern is usually implemented using a parent `State` class that contains the common functionality of all the states, and several concrete classes derived from `State`, where each derived class contains only the state-specific required functionality. The State pattern focuses on implementing a state machine. The core parts of a state machine are the states and transitions between the states. It doesn't matter how those parts are implemented.

To avoid reinventing the wheel, we can make use of existing Python modules that not only help us create state machines but also do it in a Pythonic way. A module that I find very useful is `state_machine`.

The `state_machine` module is simple enough that no special introduction is required. We will cover most aspects of it while going through the code of the example.

Let's start with the `Process` class. Each created process has its own state machine. The first step to creating a state machine using the `state_machine` module is to use the `@acts_as_state_machine` decorator. Then, we define the states of our state machine. This is a one-to-one mapping of what we see in the state diagram. The only difference is that we should give a hint about the initial state of the state machine. We do that by setting the initial attribute value to `True`:

```
@acts_as_state_machine
class Process:
```

```
    created = State(initial=True)
    waiting = State()
    running = State()
    terminated = State()
    blocked = State()
    swapped_out_waiting = State()
    swapped_out_blocked = State()
```

Next, we are going to define the transitions. In the `state_machine` module, a transition is an instance of the `Event` class. We define the possible transitions using the `from_states` and `to_state` arguments:

```
wait = Event(
    from_states=(
        created,
        running,
        blocked,
        swapped_out_waiting,
    ),
    to_state=waiting,
)
run = Event(
    from_states=waiting, to_state=running
)
terminate = Event(
    from_states=running, to_state=terminated
)
block = Event(
    from_states=(
        running,
        swapped_out_blocked,
    ),
    to_state=blocked,
)
swap_wait = Event(
    from_states=waiting,
    to_state=swapped_out_waiting,
)
swap_block = Event(
    from_states=blocked,
    to_state=swapped_out_blocked,
)
```

Also, as you may have noticed that `from_states` can be either a single state or a group of states (tuple).

Each process has a name. Officially, a process needs to have much more information to be useful (for example, ID, priority, status, and so forth) but let's keep it simple to focus on the pattern:

```python
def __init__(self, name):
    self.name = name
```

Transitions are not very useful if nothing happens when they occur. The state_machine module provides us with the @before and @after decorators that can be used to execute actions before or after a transition occurs, respectively. You can imagine updating some objects within the system or sending an email or a notification to someone. For this example, the actions are limited to printing information about the state change of the process, as follows:

```python
@after("wait")
def wait_info(self):
    print(f"{self.name} entered waiting mode")


@after("run")
def run_info(self):
    print(f"{self.name} is running")


@before("terminate")
def terminate_info(self):
    print(f"{self.name} terminated")


@after("block")
def block_info(self):
    print(f"{self.name} is blocked")


@after("swap_wait")
def swap_wait_info(self):
    print(
        f"{self.name} is swapped out and waiting"
    )


@after("swap_block")
def swap_block_info(self):
    print(
        f"{self.name} is swapped out and blocked"
    )
```

Next, we need the `transition()` function, which accepts three arguments:

- `process`, which is an instance of `Process`
- `event`, which is an instance of `Event` (wait, run, terminate, and so forth)
- `event_name`, which is the name of the event

The name of the event is printed if something goes wrong when trying to execute `event`. Here is the code for the function:

```python
def transition(proc, event, event_name):
    try:
        event()
    except InvalidStateTransition:
        msg = (
            f"Transition of {proc.name} from {proc.current_state} "
            f"to {event_name} failed"
        )
        print(msg)
```

The `state_info()` function shows some basic information about the current (active) state of the process:

```python
def state_info(proc):
    print(
        f"state of {proc.name}: {proc.current_state}"
    )
```

At the beginning of the `main()` function, we define some string constants, which are passed as `event_name`:

```python
def main():
    RUNNING = "running"
    WAITING = "waiting"
    BLOCKED = "blocked"
    TERMINATED = "terminated"
```

Next, we create two `Process` instances and display information about their initial state:

```python
p1, p2 = Process("process1"), Process(
    "process2"
)
[state_info(p) for p in (p1, p2)]
```

The rest of the function experiments with different transitions. Recall the state diagram we covered in this chapter. The allowed transitions should be with respect to the state diagram. For example, it should be possible to switch from a running state to a blocked state, but it shouldn't be possible to switch from a blocked state to a running state:

```
print()
transition(p1, p1.wait, WAITING)
transition(p2, p2.terminate, TERMINATED)
[state_info(p) for p in (p1, p2)]

print()
transition(p1, p1.run, RUNNING)
transition(p2, p2.wait, WAITING)
[state_info(p) for p in (p1, p2)]

print()
transition(p2, p2.run, RUNNING)
[state_info(p) for p in (p1, p2)]

print()
[
    transition(p, p.block, BLOCKED)
    for p in (p1, p2)
]
[state_info(p) for p in (p1, p2)]

print()
[
    transition(p, p.terminate, TERMINATED)
    for p in (p1, p2)
]
[state_info(p) for p in (p1, p2)]
```

Here is the recapitulation of the full implementation example (the ch05/state.py file):

1. We begin by importing what we need from state_machine.

2. We define the Process class with its simple attributes.

3. We add the Process class's initialization method.

4. We also need to define, in the Process class, the methods to provide its states.

5. We define the transition() function.

6. Next, we define the state_info() function.

7. Finally, we add the main function of the program.

Here's what we get when executing the Python ch05/state.py command:

```
state of process1: created
state of process2: created

process1 entered waiting mode
Transition of process2 from created to terminated failed
state of process1: waiting
state of process2: created

process1 is running
process2 entered waiting mode
state of process1: running
state of process2: waiting

process2 is running
state of process1: running
state of process2: running

process1 is blocked
process2 is blocked
state of process1: blocked
state of process2: blocked

Transition of process1 from blocked to terminated failed
Transition of process2 from blocked to terminated failed
state of process1: blocked
state of process2: blocked
```

Indeed, the output shows that illegal transitions such as created → terminated and blocked → terminated fail gracefully. We don't want the application to crash when an illegal transition is requested, and this is handled properly by the except block.

Notice how using a good module such as state_machine eliminates conditional logic. There's no need to use long and error-prone if...else statements that check for each and every state transition and react to them.

To get a better feeling for the state pattern and state machines, I strongly recommend you implement your own example. This can be anything: a simple video game (you can use state machines to handle the states of the main hero and the enemies), an elevator, a parser, or any other system that can be modeled using state machines.

The Interpreter pattern

Often, we need to create a **domain-specific language (DSL)**. A DSL is a computer language of limited expressiveness targeting a particular domain. DSLs are used for different things, such as combat simulation, billing, visualization, configuration, and communication protocols. DSLs are divided into internal DSLs and external DSLs.

Internal DSLs are built on top of a host programming language. An example of an internal DSL is a language that solves linear equations using Python. The advantages of using an internal DSL are that we don't have to worry about creating, compiling, and parsing grammar because these are already taken care of by the host language. The disadvantage is that we are constrained by the features of the host language. It is very challenging to create an expressive, concise, and fluent internal DSL if the host language does not have these features.

External DSLs do not depend on host languages. The creator of the DSL can decide all aspects of the language (grammar, syntax, and so forth). They are also responsible for creating a parser and compiler for it.

The Interpreter pattern is related only to internal DSLs. Therefore, the goal is to create a simple but useful language using the features provided by the host programming language, which in this case is Python. Note that Interpreter does not address parsing at all. It assumes that we already have the parsed data in some convenient form. This can be an **abstract syntax tree** (**AST**) or any other handy data structure [*Gang of Four-95*, page 276].

Real-world examples

A musician is an example of the Interpreter pattern. Musical notation represents the pitch and duration of a sound graphically. The musician can reproduce a sound precisely based on its notation. In a sense, musical notation is the language of music, and the musician is the interpreter of that language.

We can also cite software examples:

- In the C++ world, `boost::spirit` is considered an internal DSL for implementing parsers.
- An example in Python is PyT, an internal DSL used to generate XHTML/HTML. PyT focuses on performance and claims to have comparable speed with Jinja2. Of course, we should not assume that the Interpreter pattern is necessarily used in PyT. However, since it is an internal DSL, the Interpreter is a very good candidate for it.

Use cases for the Interpreter pattern

The Interpreter pattern is used when we want to offer a simple language to domain experts and advanced users to solve their problems. The first thing we should stress is that the Interpreter pattern should only be used to implement simple languages. If the language has the requirements of an external DSL, there are better tools to create languages from scratch (Yacc and Lex, Bison, ANTLR, and so on).

Our goal is to offer the right programming abstractions to the specialist, who is often not a programmer, to make them productive. Ideally, they shouldn't know advanced Python to use our DSL, but knowing even a little bit of Python is a plus since that's what we eventually get at the end. Advanced Python concepts should not be a requirement. Moreover, the performance of the DSL is usually not an important concern. The focus is on offering a language that hides the peculiarities of the host language and offers a more human-readable syntax. Admittedly, Python is already a very readable language with far less peculiar syntax than many other programming languages.

Implementing the Interpreter pattern

Let's create an internal DSL to control a smart house. This example fits well into the **internet of things** (**IoT**) era, which is getting more and more attention nowadays. The user can control their home using a very simple event notation. An event has the form of command -> receiver -> arguments. The arguments part is optional.

Not all events require arguments. An example of an event that does not require any arguments is shown here:

```
open -> gate
```

An example of an event that requires arguments is shown here:

```
increase -> boiler temperature -> 3 degrees
```

The - > symbol is used to mark the end of one part of an event and state the beginning of the next one. There are many ways to implement an internal DSL. We can use plain old regular expressions, string processing, a combination of operator overloading, and metaprogramming, or a library/tool that can do the hard work for us. Although, officially, the Interpreter pattern does not address parsing, I feel that a practical example needs to cover parsing as well. For this reason, I decided to use a tool to take care of the parsing part. The tool is called pyparsing and, to find out more about it, check out the mini-book *Getting Started with Pyparsing* by Paul McGuire (`https://www.oreilly.com/library/view/getting-started-with/9780596514235/`).

Before getting into coding, it is a good practice to define a simple grammar for our language. We can define the grammar using the **Backus-Naur Form** (**BNF**) notation:

```
event ::= command token receiver token arguments
command ::= word+
word ::= a collection of one or more alphanumeric characters
token ::= ->
receiver ::= word+
arguments ::= word+
```

What the grammar basically tells us is that an event has the form of command -> receiver -> arguments, and that commands, receivers, and arguments have the same form: a group of one or more alphanumeric characters. If you are wondering about the necessity of the numeric part, it is included to allow us to pass arguments, such as three degrees at the increase -> boiler temperature -> 3 degrees command.

Now that we have defined the grammar, we can move on to converting it to actual code. Here's what the code looks like:

```
word = Word(alphanums)
command = Group(OneOrMore(word))
token = Suppre"s("->")
device = Group(OneOrMore(word))
argument = Group(OneOrMore(word))
event = command + token + device + Optional(token + argument)
```

The basic difference between the code and grammar definition is that the code needs to be written in the bottom-up approach. For instance, we cannot use a word without first assigning it a value. Suppress is used to state that we want the - > symbol to be skipped from the parsed results.

The full code of the final implementation example (see the ch05/interpreter/interpreter. py file) uses many placeholder classes, but to keep you focused, I will first show a minimal version featuring only one class. Let's look at the Boiler class. A boiler has a default temperature of 83° Celsius. There are also two methods to increase and decrease the current temperature:

```
class Boiler:
    def __init__(self):
        self.temperature = 83   # in celsius

    def __str__(self):
        return f"boiler temperature: {self.temperature}"

    def increase_temperature(self, amount):
        print(f"increasing the boiler's temperature by {amount}
degrees")
        self.temperature += amount

    def decrease_temperature(self, amount):
        print(f"decreasing the boiler's temperature by {amount}
degrees")
        self.temperature -= amount
```

The next step is to add the grammar, which we already covered. We will also create a boiler instance and print its default state:

```
word = Word(alphanums)
command = Group(OneOrMore(word))
```

```
token = Suppress("->")
device = Group(OneOrMore(word))
argument = Group(OneOrMore(word))
event = command + token + device + Optional(token + argument)

boiler = Boiler()
```

The simplest way to retrieve the parsed output of pyparsing is by using the `parseString()` method. The result is a `ParseResults` instance, which is a parse tree that can be treated as a nested list. For example, executing `print(event.parseStri'g('increase -> boiler temperature -> 3 degr'es'))` would give `[['incre'se']' ['boi'er', 'temperat're']' ''3', 'degr'es']]` as a result.

So, in this case, we know that the first sublist is the *command* (increase), the second sublist is the *receiver* (boiler temperature), and the third sublist is the *argument* (3°). We can unpack the `ParseResults` instance, which gives us direct access to these three parts of the event. Having direct access means that we can match patterns to find out which method should be executed:

```
test = "increase -> boiler temperature -> 3 degrees"
cmd, dev, arg = event.parseString(test)
cmd_str = " ".join(cmd)
dev_str = " ".join(dev)

if "increase" in cmd_str and "boiler" in dev_str:
    boiler.increase_temperature(int(arg[0]))

print(boiler)
```

Executing the preceding code snippet (using `python ch05/interpreter/boiler.py`) gives the following output:

```
increasing the boiler's temperature by 3 degrees
boiler temperature: 86
```

The full code for our implementation (in the `ch05/interpreter/interpreter.py` file) is not very different from what I just described. It is just extended to support more events and devices. Let's summarize the steps here:

1. First, we import all we need from `pyparsing`.

2. We define the following classes: `Gate`, `Aircondition`, `Heating`, `Boiler` (already presented), and `Fridge`.

3. Next, we have our main function:

    A. We prepare the parameters for the tests we will be performing, using the following variables: `tests`, `open_actions`, and `close_actions`.

    B. We execute the test actions.

Executing the `python ch05/interpreter/interpreter.py` command gives the following output:

```
opening the gate
closing the garage
turning on the air condition
turning off the heating
increasing the boiler's temperature by 5 degrees
decreasing the fridge's temperature by 2 degrees
```

If you want to experiment more with this example, I have a few suggestions for you. The first change that will make it much more interesting is to make it interactive. Currently, all the events are hardcoded in the `tests` tuple. However, the user wants to be able to activate events using an interactive prompt. Do not forget to check how sensitive `pyparsing` is regarding spaces, tabs, or unexpected input. For example, what happens if the user types `turn off -> heating 37`?

The Strategy pattern

Several solutions often exist for the same problem. Consider the task of sorting, which involves arranging the elements of a list in a particular sequence. For example, a variety of sorting algorithms are available for the task of sorting. Generally, no single algorithm outperforms all others in every situation.

Selecting a sorting algorithm depends on various factors, tailored to the specifics of each case. Some key considerations include the following:

- **The number of elements to be sorted, known as the input size**: While most sorting algorithms perform adequately with a small input size, only a select few maintain efficiency with larger datasets.

- **The best/average/worst time complexity of the algorithm**: Time complexity is (roughly) the amount of time the algorithm takes to complete, excluding coefficients and lower-order terms. This is often the most usual criterion to pick an algorithm, although it is not always sufficient.

- **The space complexity of the algorithm**: Space complexity is (again roughly) the amount of physical memory needed to fully execute an algorithm. This is very important when we are working with big data or embedded systems, which usually have limited memory.

- **Stability of the algorithm**: An algorithm is considered stable when it maintains the relative order of elements with equal values after it is executed.

- **Code complexity of the algorithm**: If two algorithms have the same time/space complexity and are both stable, it is important to know which algorithm is easier to code and maintain.

Other factors might also influence the choice of a sorting algorithm. The key consideration is whether a single algorithm must be applied universally. The answer is, unsurprisingly, no. It is more practical to have access to various sorting algorithms and choose the most suitable one for a given situation, based on the criteria. That's what the Strategy pattern is about.

The Strategy pattern promotes using multiple algorithms to solve a problem. Its killer feature is that it makes it possible to switch algorithms at runtime transparently (the client code is unaware of the change). So, if you have two algorithms and you know that one works better with small input sizes, while the other works better with large input sizes, you can use Strategy to decide which algorithm to use based on the input data at runtime.

Real-world examples

Reaching an airport to catch a flight is a good real-life Strategy example:

- If we want to save money and we leave early, we can go by bus/train
- If we don't mind paying for a parking place and have our own car, we can go by car
- If we don't have a car but we are in a hurry, we can take a taxi

There are trade-offs between cost, time, convenience, and so forth.

In software, Python's `sorted()` and `list.sort()` functions are examples of the Strategy pattern. Both functions accept a named parameter key, which is basically the name of the function that implements a sorting strategy (*Python 3 Patterns, Recipes, and Idioms, by Bruce Eckel & Friends*, page 202).

Use cases for the Strategy pattern

Strategy is a very generic design pattern with many use cases. In general, whenever we want to be able to apply different algorithms dynamically and transparently, Strategy is the way to go. By different algorithms, I mean different implementations of the same algorithm. This means that the result should be the same, but each implementation has a different performance and code complexity (as an example, think of sequential search versus binary search).

Apart from its usage for sorting algorithms as we mentioned, the Strategy pattern is used to create different formatting representations, either to achieve portability (for example, line-breaking differences between platforms) or dynamically change the representation of data.

Implementing the Strategy pattern

There is not much to be said about implementing the Strategy pattern. In languages where functions are not first-class citizens, each Strategy should be implemented in a different class. In Python, functions are objects (we can use variables to reference and manipulate them) and this simplifies the implementation of Strategy.

Assume that we are asked to implement an algorithm to check whether all characters in a string are unique. For example, the algorithm should return true if we enter the dream string because none of the characters are repeated. If we enter the pizza string, it should return false because the letter "z" exists two times. Note that the repeated characters do not need to be consecutive, and the string does not need to be a valid word. The algorithm should also return false for the 1r2a3ae string because the letter "a" appears twice.

After thinking about the problem carefully, we come up with an implementation that sorts the string and compares all characters pair by pair. First, we implement the `pairs()` function, which returns all neighbors pairs of a sequence, `seq`:

```
def pairs(seq):
    n = len(seq)
    for i in range(n):
        yield seq[i], seq[(i + 1) % n]
```

Next, we implement the `allUniqueSort()` function, which accepts a string, s, and returns `True` if all characters in the string are unique; otherwise, it returns `False`. To demonstrate the Strategy pattern, we will simplify by assuming that this algorithm fails to scale. We assume that it works fine for strings that are up to five characters. For longer strings, we simulate a slowdown by inserting a sleep statement:

```
SLOW = 3  # in seconds
LIMIT = 5  # in characters
WARNING"= "too bad, you picked the slow algorithm":("


def allUniqueSort(s):
    if len(s) > LIMIT:
        print(WARNING)
        time.sleep(SLOW)
    srtStr = sorted(s)
    for c1, c2 in pairs(srtStr):
        if c1 == c2:
            return False
    return True
```

We are not happy with the performance of `allUniqueSort()`, and we are trying to think of ways to improve it. After some time, we come up with a new algorithm, `allUniqueSet()`, that eliminates the need to sort. In this case, we use a set. If the character in check has already been inserted in the set, it means that not all characters in the string are unique:

```
def allUniqueSet(s):
    if len(s) < LIMIT:
        print(WARNING)
```

```
        time.sleep(SLOW)

    return True if len(set(s)) == len(s) else False
```

Unfortunately, while allUniqueSet() has no scaling problems, for some strange reason, it performs worse than allUniqueSort() when checking short strings. What can we do in this case? Well, we can keep both algorithms and use the one that fits best, depending on the length of the string that we want to check.

The allUnique() function accepts an input string, s, and a strategy function, strategy, which, in this case, is one of allUniqueSort() and allUniqueSet(). The allUnique() function executes the input strategy and returns its result to the caller.

Then, the main() function lets the user perform the following actions:

- Enter the word to be checked for character uniqueness
- Choose the pattern that will be used

It also does some basic error handling and gives the ability to the user to quit gracefully:

```
def main():
    WORD_IN_DESC = "Insert word (type quit to exit)> "
    STRAT_IN_DESC = "Choose strategy: [1] Use a set, [2] Sort and
pair> "

    while True:
        word = None
        while not word:
            word = input(WORD_IN_DESC)

        if word == "quit":
            print("bye")
            return

        strategy_picked = None
        strategies = {"1": allUniqueSet, "2": allUniqueSort}
        while strategy_picked not in strategies.keys():
            strategy_picked = input(STRAT_IN_DESC)

        try:
            strategy = strategies[strategy_picked]
            result = allUnique(word, strategy)
            print(f"allUnique({word}): {result}")
        except KeyError:
            print(f"Incorrect option: {strategy_picked}")
```

Here's a summary of the complete code for our implementation example (the ch05/strategy. py file):

1. We import the time module.

2. We define the pairs() function.

3. We define the values for the SLOW, LIMIT, and WARNING constants.

4. We define the function for the first algorithm, allUniqueSort().

5. We define the function for the second algorithm, allUniqueSet().

6. Next, we define the allUnique() function that helps call a chosen algorithm by passing the corresponding strategy function.

7. Finally, we add the main() function.

Let's see the output of a sample execution using the python ch05/strategy.py command:

```
Insert word (type quit to exit)> balloon
Choose strategy: [1] Use a set, [2] Sort and pair> 1
allUnique(balloon): False
Insert word (type quit to exit)> balloon
Choose strategy: [1] Use a set, [2] Sort and pair> 2
too bad, you picked the slow algorithm :(
allUnique(balloon): False
Insert word (type quit to exit)> bye
Choose strategy: [1] Use a set, [2] Sort and pair> 1
too bad, you picked the slow algorithm :(
allUnique(bye): True
Insert word (type quit to exit)> bye
Choose strategy: [1] Use a set, [2] Sort and pair> 2
allUnique(bye): True
Insert word (type quit to exit)>
```

The first word, balloon, has more than five characters and not all of them are unique. In this case, both algorithms return the correct result, False, but allUniqueSort() is slower and the user is warned.

The second word, bye, has less than five characters and all characters are unique. Again, both algorithms return the expected result, True, but this time, allUniqueSet() is slower and the user is warned once more.

Normally, the strategy that we want to use should not be picked by the user. The point of the strategy pattern is that it makes it possible to use different algorithms transparently. Change the code so that the faster algorithm is always picked.

There are two usual users of our code. One is the end user, who should be unaware of what's happening in the code, and to achieve that we can follow the tips given in the previous paragraph. Another possible category of users is the other developers. Assume that we want to create an API that will be used by the other developers. How can we keep them unaware of the Strategy pattern? A tip is to think of encapsulating the two functions in a common class, for example, `AllUnique`. In this case, the other developers will just need to create an instance of that class and execute a single method, for instance, `test()`. What needs to be done in this method?

The Memento pattern

In many situations, we need a way to easily take a snapshot of the internal state of an object, so that we can restore the object with it when needed. Memento is a design pattern that can help us implement a solution for such situations.

The Memento design pattern has three key components:

- **Memento**: A simple object that contains basic state storage and retrieval capabilities
- **Originator**: An object that gets and sets values of Memento instances
- **Caretaker**: An object that can store and retrieve all previously created Memento instances

Memento shares many similarities with the Command pattern.

Real-world examples

The Memento pattern can be seen in many situations in real life.

An example could be found in the dictionary we use for a language, such as English or French. The dictionary is regularly updated through the work of academic experts, with new words being added and other words becoming obsolete. Spoken and written languages evolve, and the official dictionary has to reflect that. From time to time, we revisit a previous edition to get an understanding of how the language was used at some point in the past. This could also be needed simply because information can be lost after a long period of time, and to find it, you may need to look into old editions. This can be useful for understanding something in a particular field. Someone doing research could use an old dictionary or go to the archives to find information about some words and expressions.

This example can be extended to other written material, such as books and newspapers.

Zope (`http://www.zope.org`), with its integrated object database called **Zope Object Database (ZODB)**, offers a good software example of the Memento pattern. It is famous for its **Through-The-Web** object management interface, with undo support, for website administrators. ZODB is an object database for Python and is in heavy use in the Pyramid and Plone stacks among others.

Use cases for the Memento pattern

Memento is usually used when you need to provide some sort of undo and redo capability for your users.

Another usage is the implementation of a UI dialog with OK/Cancel buttons, where we would store the state of the object on load, and if the user chooses to cancel, we would restore the initial state of the object.

Implementing the Memento pattern

We will approach the implementation of Memento, in a simplified way, and by doing things in a natural way for the Python language. This means we do not necessarily need several classes.

One thing we will use is Python's `pickle` module. What is `pickle` used for? According to the module's documentation (`https://docs.python.org/3/library/pickle.html`), the `pickle` module can transform a complex object into a byte stream, and it can transform the byte stream into an object with the same internal structure.

> **Warning**
>
> The `pickle` module is used here for the sake of our demonstration, but you should know that it is not secure for generic usage.

Let's take a `Quote` class, with the `text` and `author` attributes. To create memento, we will use a method on that class, `save_state()`, which as the name suggests will dump the state of the object, using the `pickle.dumps()` function. This creates memento:

```python
class Quote:
    def __init__(self, text, author):
        self.text = text
        self.author = author

    def save_state(self):
        current_state = pickle.dumps(self.__dict__)

        return current_state
```

That state can be restored later. For that, we add the `restore_state()` method, making use of the `pickle.loads()` function:

```python
    def restore_state(self, memento):
        previous_state = pickle.loads(memento)

        self.__dict__.clear()
        self.__dict__.update(previous_state)
```

Let's also add the `__str__` method:

```python
def __str__(self):
    return f"{self.text}\n- By {self.author}."
```

Then, in the main function, we can take care of things and test our implementation, as usual:

```python
def main():
    print("** Quote 1 **")
    q1 = Quote(
        "A room without books is like a body without a soul.",
        "Unknown author",
    )
    print(f"\nOriginal version:\n{q1}")
    q1_mem = q1.save_state()

    # Now, we found the author's name
    q1.author = "Marcus Tullius Cicero"
    print(f"\nWe found the author, and did an updated:\n{q1}")

    # Restoring previous state (Undo)
    q1.restore_state(q1_mem)
    print(f"\nWe had to restore the previous version:\n{q1}")

    print()
    print("** Quote 2 **")
    text = (
        "To be you in a world that is constantly \n"
        "trying to make you be something else is \n"
        "the greatest accomplishment."
    )
    q2 = Quote(
        text,
        "Ralph Waldo Emerson",
    )
    print(f"\nOriginal version:\n{q2}")
    _ = q2.save_state()

    # changes to the text
    q2.text = (
        "To be yourself in a world that is constantly \n"
        "trying to make you something else is the greatest \n"
        "accomplishment."
    )
```

```
    print(f"\nWe fixed the text:\n{q2}")
    q2_mem2 = q2.save_state()

    q2.text = (
        "To be yourself when the world is constantly \n"
        "trying to make you something else is the greatest \n"
        "accomplishment."
    )
    print(f"\nWe fixed the text again:\n{q2}")

    # Restoring previous state (Undo)
    q2.restore_state(q2_mem2)
    print(f"\nWe restored the 2nd version, the correct one:\n{q2}")
```

Here's the recapitulation of the steps in the example (the ch05/memento.py file):

1. We import the `pickle` module.

2. We define the `Quote` class.

3. Finally, we add the main function where we test the implementation.

Let's view a sample execution using the `python ch05/memento.py` command:

```
** Quote 1 **

Original version:
A room without books is like a body without a soul.
- By Unknown author.

We found the author, and did an updated:
A room without books is like a body without a soul.
- By Marcus Tullius Cicero.

We had to restore the previous version:
A room without books is like a body without a soul.
- By Unknown author.

** Quote 2 **

Original version:
To be you in a world that is constantly
trying to make you be something else is
the greatest accomplishment.
- By Ralph Waldo Emerson.
```

```
We fixed the text:
To be yourself in a world that is constantly
trying to make you something else is the greatest
accomplishment.
- By Ralph Waldo Emerson.

We fixed the text again:
To be yourself when the world is constantly
trying to make you something else is the greatest
accomplishment.
- By Ralph Waldo Emerson.

We restored the 2nd version, the correct one:
To be yourself in a world that is constantly
trying to make you something else is the greatest
accomplishment.
- By Ralph Waldo Emerson.
```

The output shows the program does what we expected: we can restore a previous state for each of our Quote objects.

The Iterator pattern

In programming, we use sequences or collections of objects a lot, particularly in algorithms and when writing programs that manipulate data. One can think of automation scripts, APIs, data-driven apps, and other domains. In this chapter, we are going to see a pattern that is useful whenever we must handle collections of objects: the Iterator pattern.

> **Note, according to the definition given by Wikipedia**
>
> *Iterator is a design pattern in which an iterator is used to traverse a container and access the container's elements. The iterator pattern decouples algorithms from containers; in some cases, algorithms are necessarily container-specific and thus cannot be decoupled.*

The Iterator pattern is extensively used in the Python context. As we will see, this translates into Iterator being a language feature. It is so useful that the language developers decided to make it a feature.

Use cases for the Iterator pattern

It is a good idea to use the Iterator pattern whenever you want one or several of the following behaviors:

- Make it easy to navigate through a collection

- Get the next object in the collection at any point

- Stop when you are done traversing through the collection

Implementing the Iterator pattern

Iterator is implemented in Python for us, within for loops, list comprehensions, and so on. Iterator in Python is simply an object that can be iterated upon; an object that will return data, one element at a time.

We can do our own implementation for special cases, using the Iterator protocol, meaning that our iterator object must implement two special methods: __iter__() and __next__().

An object is called iterable if we can get an iterator from it. Most of the built-in containers in Python (list, tuple, set, string, and so on) are iterable. The iter() function (which in turn calls the __iter__() method) returns an iterator from them.

Let's consider a football team we want to implement with the help of the FootballTeam class. If we want to make an iterator out of it, we have to implement the Iterator protocol, since it is not a built-in container type such as the list type. Basically, built-in iter() and next() functions would not work on it unless they are added to the implementation.

First, we define the class of the iterator, FootballTeamIterator, that will be used to iterate through the football team object. The members attribute allows us to initialize the iterator object with our container object (which will be a FootballTeam instance). We add a __iter__() method to it, which would return the object itself, and a __next__() method to return the next person from the team at each call until we reach the last person. These will allow looping over the members of the football team via the iterator. The whole code for the FootballTeamIterator class is as follows:

```python
class FootballTeamIterator:
    def __init__(self, members):
        self.members = members
        self.index = 0

    def __iter__(self):
        return self

    def __next__(self):
        if self.index < len(self.members):
            val = self.members[self.index]
            self.index += 1
            return val
        else:
            raise StopIteration()
```

So, now for the FootballTeam class itself; the next thing to do is add a __iter__() method to it, which will initialize the iterator object that it needs (thus using FootballTeamIterator(self.members)) and return it:

```
class FootballTeam:
    def __init__(self, members):
        self.members = members

    def __iter__(self):
        return FootballTeamIterator(self.members)
```

We add a small main function to test our implementation. Once we have a FootballTeam instance, we call the iter() function on it to create the iterator, and we loop through it using a while loop:

```
def main():
    members = [f"player{str(x)}" for x in range(1, 23)]
    members = members + ["coach1", "coach2", "coach3"]
    team = FootballTeam(members)
    team_it = iter(team)

    try:
        while True:
            print(next(team_it))
    except StopIteration:
        print("(End)")
```

Here is a recap of the steps in our example (the ch05/iterator.py file):

1. We define the class for the iterator.
2. We define the container class.
3. We define our main function followed by the snippet to call it.

Here is the output we get when executing the python ch05/iterator.py command:

```
player1
player2
player3
player4
player5
...
player22
coach1
coach2
coach3
(End)
```

We got the expected output. Also, we can see that an exception was raised when we reached the end of the iteration, but it was caught, printing the (End) string instead.

The Template pattern

A key ingredient in writing good code is avoiding redundancy. In OOP, methods and functions are important tools that we can use to avoid writing redundant code.

Remember the sorted() example we saw when discussing the Strategy pattern. That function is generic enough that it can be used to sort more than one data structure (lists, tuples, and named tuples) using arbitrary keys. That's the definition of a good function.

Functions such as sorted() demonstrate the ideal case. However, we cannot always write 100% generic code.

In the process of writing code that handles algorithms in the real world, we often end up writing redundant code. That's the problem solved by the Template design pattern. This pattern focuses on eliminating code redundancy. The idea is that we should be able to redefine certain parts of an algorithm without changing its structure.

Real-world examples

The daily routine of a worker, especially for workers of the same company, is very close to the Template design pattern. All workers follow the same routine, but specific parts of the routine are very different.

In software, Python uses the Template pattern in the cmd module, which is used to build line-oriented command interpreters. Specifically, cmd.Cmd.cmdloop() implements an algorithm that reads input commands continuously and dispatches them to action methods. What is done before the loop, after the loop, and at the command parsing part is always the same. This is also called the **invariant** part of an algorithm. The elements that change are the actual action methods (the variant part).

Use cases for the Template pattern

The Template design pattern focuses on eliminating code repetition. If we notice that there is repeatable code in algorithms that have structural similarities, we can keep the invariant (common) parts of the algorithms in a Template method/function and move the variant (different) parts in action/hook methods/functions.

Pagination is a good use case to use Template. A pagination algorithm can be split into an abstract (invariant) part and a concrete (variant) part. The invariant part takes care of things such as the maximum number of lines/pages. The variant part contains functionality to show the header and footer of a specific page that is paginated.

All application frameworks make use of some form of the Template pattern. When we use a framework to create a graphical application, we usually inherit from a class and implement our custom behavior. However, before this, a Template method is usually called, which implements the part of the application that is always the same, which is drawing the screen, handling the event loop, resizing and centralizing the window, and so on (*Python 3 Patterns, Recipes and Idioms, by Bruce Eckel & Friends*, page 133).

Implementing the Template pattern

In this example, we will implement a banner generator. The idea is rather simple. We want to send some text to a function, and the function should generate a banner containing the text. Banners have some sort of style, for example, dots or dashes surrounding the text. The banner generator has a default style, but we should be able to provide our own style.

The generate_banner() function is our Template function. It accepts, as an input, the text (msg) that we want our banner to contain, and the style (style) that we want to use. The generate_ banner() function wraps the styled text with a simple header and footer. The header and footer can be much more complex, but nothing forbids us from calling functions that can do the header and footer generations instead of just printing simple strings:

```python
def generate_banner(msg, style):
    print("-- start of banner --")
    print(style(msg))
    print("-- end of banner --nn")
```

The dots_style() function simply capitalizes msg and prints 10 dots before and after it:

```python
def dots_style(msg):
    msg = msg.capitalize()
    ten_dots = "." * 10
    msg = f"{ten_dots}{msg}{ten_dots}"
    return msg
```

Another style that is supported by the generator is admire_style(). This style shows the text in uppercase and puts an exclamation mark between each character of the text:

```python
def admire_style(msg):
    msg = msg.upper()
    return "!".join(msg)
```

The next style is by far my favorite. The cow_style() style executes the milk_random_cow() method of cowpy, which is used to generate a random ASCII art character every time cow_style() is executed. Here is the cow_style() function:

```python
def cow_style(msg):
    msg = cow.milk_random_cow(msg)
    return msg
```

The main() function sends the "happy coding" text to the banner and prints it to the standard output using all the available styles:

```
def main():
    styles = (dots_style, admire_style, cow_style)
    msg = "happy coding"
    [generate_banner(msg, style) for style in styles]
```

Here is the recap of the full code of the example (the ch05/template.py file):

1. We import the cow function from cowpy.

2. We define the generate_banner() function.

3. We define the dots_style() function.

4. Next, we define the admire_style() and cow_style() functions.

5. We finish with the main function and the snippet to call it.

Let's look at a sample output by executing python ch05/template.py (note that your cow_style() output might be different due to the randomness of cowpy):

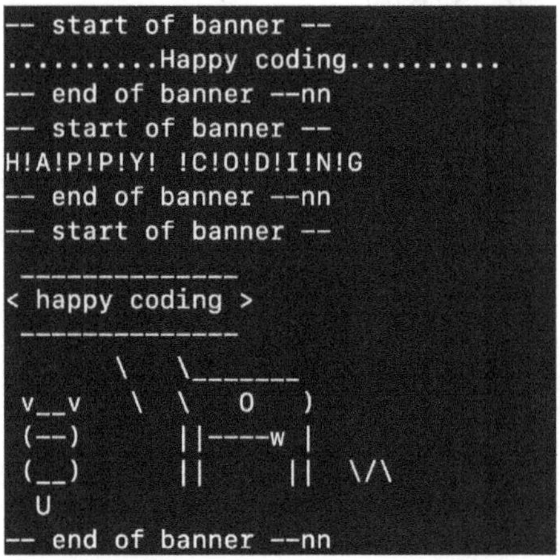

Figure 5.3 – Sample art output of the ch05/template.py program

Do you like the art generated by cowpy? I certainly do. As an exercise, you can create your own style and add it to the banner generator.

Another good exercise is to try implementing your own *Template* example. Find some existing redundant code that you wrote and see whether this pattern is applicable.

Other behavioral design patterns

What about the other behavioral design patterns from the Gang of Four's catalog? We also have the **Mediator pattern** and the **Visitor pattern**:

- The Mediator pattern promotes loose coupling between objects by encapsulating how they interact and communicate with each other. In this pattern, objects don't communicate directly with each other; instead, they communicate through a mediator object. This mediator object acts as a central hub that coordinates communication between the objects. The Mediator pattern stands out as a solution for promoting loose coupling and managing complex interactions between objects.

- For complex use cases, the Visitor pattern provides a solution for separating algorithms from the objects on which they operate. By allowing new operations to be defined without modifying the classes of the elements on which they operate, the Visitor pattern promotes flexibility and extensibility in object-oriented systems.

We are not going to discuss these two patterns, since they are not commonly used by Python developers. Python offers built-in features and libraries that can help achieve loose coupling and/or extensibility goals without the need to implement these patterns. For example, one can use event-driven programming with a library such as `asyncio` instead of communication between objects through a mediator object. Additionally, using functions as first-class citizens, decorators, or context managers can provide ways to encapsulate algorithms and operations without the need for explicit visitor objects.

Summary

In this chapter, we discussed the behavioral design patterns.

First, we covered the Chain of Responsibility pattern, which simplifies the management of complex processing flows, making it a valuable tool for enhancing flexibility and maintainability in software design.

Second, we went over the Command pattern, which encapsulates a request as an object, thereby allowing us to parameterize clients with queues, requests, and operations. It also allows us to support undoable operations. Although the most advertised feature of command by far is undo, it has more uses. In general, any operation that can be executed at the user's will at runtime is a good candidate for using the Command pattern.

We looked at the Observer pattern, which helps with the separation of concerns, increasing decoupling between the publisher and subscribers. We have seen that observers are loosely coupled with the subject and can be added or removed dynamically.

Then, we went over the State pattern, which is an implementation of one or more state machines used to solve a particular software engineering problem. A state machine can have only one active state at any point in time. A transition is a switch from the current state to a new state. It is normal to execute one or more actions before or after a transition occurs. State machines can be represented visually using state diagrams. State machines are used to solve many computational and non-computational problems. We saw how to implement a state machine for a computer system process using the `state_machine` module. The `state_machine` module simplifies the creation of a state machine and the definition of actions before/after transitions.

Afterward, we looked at the Interpreter pattern, which is used to offer a programming-like framework to advanced users and domain experts, without exposing the complexities of a programming language. This is achieved by implementing a DSL, a computer language that has limited expressiveness and targets a specific domain. The interpreter is related to what are called internal DSLs. Although parsing is generally not addressed by the Interpreter pattern, as an implementation example, we used pyparsing to create a DSL that controls a smart house and saw that using a good parsing tool makes interpreting the results using pattern matching simple.

Then, we looked at the Strategy design pattern, which is generally used when we want to be able to use multiple solutions for the same problem, transparently. There is no perfect algorithm for all input data and all cases, and by using Strategy, we can dynamically decide which algorithm to use in each case. We saw how Python, with its first-class functions, simplifies the implementation of Strategy by implementing two different algorithms that check whether all of the characters in a word are unique.

Next, we looked at the Memento pattern, which is used to store the state of an object when needed. Memento provides an efficient solution when implementing some sort of undo capability for your users. Another usage is the implementation of a UI dialog with OK/Cancel buttons, where, if the user chooses to cancel, we will restore the initial state of the object. We used an example to get a feel for how Memento, in a simplified form and using the `pickle` module from the standard library, can be used in an implementation where we want to be able to restore previous states of data objects.

We then looked at the Iterator pattern, which gives a nice and efficient way to iterate through sequences and collections of objects. In real life, whenever you have a collection of things and you are getting to those things one by one, you are using a form of the Iterator pattern. In Python, Iterator is a language feature. We can use it immediately on built-in containers such as lists and dictionaries, and we can define new iterable and iterator classes, to solve our problem, by using the Python iterator protocol. We saw that with an example of implementing a football team.

Then, we saw how we can use the Template pattern to eliminate redundant code when implementing algorithms with structural similarities. We saw how the daily routine of a worker resembles the Template pattern. We also mentioned two examples of how Python uses Template in its libraries. General use cases of when to use Template were also mentioned. We concluded by implementing a banner generator, which uses a Template function to implement custom text styles.

There are other structural design patterns: Mediator and Visitor. They are not commonly used by Python developers; therefore, we have not discussed them.

In the next chapter, we will explore architectural design patterns, which are patterns that help in solving common architectural problems.

Part 3:
Beyond the Gang of Four

This part goes beyond the classic design patterns to help you address special software design needs such as microservices, cloud-based applications, and performance optimization. It also discusses patterns for testing and specific Python anti-patterns

- *Chapter 6, Architectural Design Patterns*

- *Chapter 7, Concurrency and Asynchronous Patterns*

- *Chapter 8, Performance Patterns*

- *Chapter 9, Distributed Systems Patterns*

- *Chapter 10, Patterns for Testing*

- *Chapter 11, Python Anti-Patterns*

6
Architectural Design Patterns

In the previous chapter, we covered **behavioral patterns**, patterns that help with object interconnection and algorithms. The next category of design patterns is **architectural design patterns**. These patterns provide a template for solving common architectural problems, facilitating the development of scalable, maintainable, and reusable systems.

In this chapter, we're going to cover the following main topics:

- The **Model-View-Controller** (**MVC**) pattern
- The **Microservices** pattern
- The **Serverless** pattern
- The **Event Sourcing** pattern
- Other architectural design patterns

At the end of this chapter, you will understand how to build robust and flexible software using popular architectural design patterns.

Technical requirements

See the requirements presented in *Chapter 1*. The additional technical requirements for the code discussed in this chapter are the following:

- For the *Microservices pattern* section, install the following:

  - **gRPC**, using the following command: `python -m pip install grpcio`

  - **gRPC-tools**, using the following command: `python -m pip install grpcio-tools`

  - **Lanarky** and its dependencies, using the following command: `python -m pip install "lanarky[openai]"==0.8.6 uvicorn==0.29.0` (Note that this is not compatible with Python 3.12, at the time of writing. In this case, you may reproduce the related example using Python 3.11 instead.)

- For the *Serverless pattern* section, install the following:

  - **Docker**
  - **LocalStack**, for testing AWS Lambda locally, using the following command: `python -m pip install localstack` (Note that this is not compatible with Python 3.12, at the time of writing. You may use Python 3.11 instead for this case.)
  - **awscli-local**, using the command: `python -m pip install awscli-local`
  - `awscli`, using the command: `python -m pip install awscli`

- For the *Event Sourcing* section, install the following:

  - `eventsourcing`, using the command: `python -m pip install eventsourcing`

The MVC pattern

The MVC pattern is another application of the **loose coupling** principle. The name of the pattern comes from the three main components used to split a software application: the model, the view, and the controller.

Even if we will never have to implement it from scratch, we need to be familiar with it because all common frameworks use MVC or a slightly different version of it (more on this later).

The model is the core component. It represents knowledge. It contains and manages the (business) logic, data, state, and rules of an application. The view is a visual representation of the model. Examples of views are a computer GUI, the text output of a computer terminal, a smartphone's application GUI, a PDF document, a pie chart, a bar chart, and so forth. The view only displays the data; it doesn't handle it. The controller is the link/glue between the model and the view. All communication between the model and the view happens through a controller.

A typical use of an application that uses MVC, after the initial screen is rendered to the user, is as follows:

1. The user triggers a view by clicking (typing, touching, and so on) a button.
2. The view informs the controller of the user's action.
3. The controller processes user input and interacts with the model.
4. The model performs all the necessary validation and state changes and informs the controller about what should be done.
5. The controller instructs the view to update and display the output appropriately, following the model's instructions.

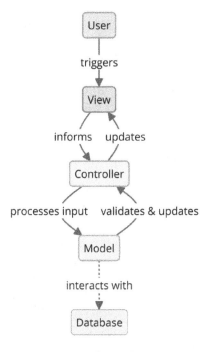

Figure 6.1 – The MVC pattern

But is the controller part necessary? Can't we just skip it? We could, but then we would lose a big benefit that MVC provides: the ability to use more than one view (even at the same time, if that's what we want) without modifying the model. To achieve decoupling between the model and its representation, every view typically needs its own controller. If the model communicated directly with a specific view, we wouldn't be able to use multiple views (or, at least, not in a clean and modular way).

Real-world examples

MVC is an application of the **separation of concern** principle. Separation of concern is used a lot in real life. For example, if you build a new house, you usually assign different professionals to 1) install the plumbing and electricity; and, 2) paint the house.

Another example is a restaurant. In a restaurant, the waiters receive orders and serve dishes to the customers, but the meals are cooked by the chefs.

In web development, several frameworks use the MVC idea, for example:

- The **Web2py framework** is a lightweight Python framework that embraces the MVC pattern. There are many examples that demonstrate how MVC can be used in Web2py on the project's site (http://web2py.com/) and in the GitHub repository.

- **Django** (`https://www.djangoproject.com/`) is also an MVC framework, although it uses different naming conventions. The controller is called *view*, and the view is called *template*. Django uses the name **Model-View Template (MVT)**. According to the designers of Django, the view describes what data is seen by the user, and therefore, it uses the name view as the Python callback function for a particular URL. The term "template" in Django is used to separate content from representation. It describes how the data is seen by the user, not which data is seen.

Use cases for the MVC pattern

MVC is a very generic and useful design pattern. In fact, all popular web frameworks (Django, Rails, Symfony, and Yii) and application frameworks (iPhone SDK, Android, and QT) make use of MVC or a variation of it (**model-view-adapter (MVA)**, **model-view-presenter (MVP)**, or **MVT**, for example). However, even if we don't use any of these frameworks, it makes sense to implement the pattern on our own because of the benefits it provides, which are as follows:

- The separation between the view and model allows graphics designers to focus on the **user interface (UI)** part and programmers to focus on development, without interfering with each other.

- Because of the loose coupling between the view and model, each part can be modified/extended without affecting the other. For example, adding a new view is trivial. Just implement a new controller for it.

- Maintaining each part is easier because the responsibilities are clear.

When implementing MVC from scratch, be sure that you create smart models, thin controllers, and dumb views.

A model is considered smart because it does the following:

- It contains all the validation/business rules/logic

- It handles the state of the application

- It has access to application data (database, cloud, and so on)

- It does not depend on the UI

A controller is considered thin because it does the following:

- It updates the model when the user interacts with the view

- It updates the view when the model changes

- It processes the data before delivering it to the model/view, if necessary

- It does not display the data

- It does not access the application data directly

- It does not contain validation/business rules/logic

A view is considered dumb because it does the following:

- It displays the data

- It allows the user to interact with it

- It does only minimal processing, usually provided by a template language (for example, using simple variables and loop controls)

- It does not store any data

- It does not access the application data directly

- It does not contain validation/business rules/logic

If you are implementing MVC from scratch and want to find out whether you did it right, you can try answering some key questions:

- If your application has a GUI, is it skinnable? How easily can you change the skin/look and feel of it? Can you give the user the ability to change the skin of your application during runtime? If this is not simple, it means that something is going wrong with your MVC implementation.

- If your application has no GUI (for instance, if it's a terminal application), how hard is it to add GUI support? Or, if adding a GUI is irrelevant, is it easy to add views to display the results in a chart (pie chart, bar chart, and so on) or a document (PDF, spreadsheet, and so on)? If these changes are not trivial (a matter of creating a new controller with a view attached to it, without modifying the model), MVC is not implemented properly.

- If you make sure that these conditions are satisfied, your application will be more flexible and maintainable compared to an application that does not use MVC.

Implementing the MVC pattern

I could use any of the common frameworks to demonstrate how to use MVC, but I feel that the picture will be incomplete. So, I decided to show you how to implement MVC from scratch, using a very simple example: a quote printer. The idea is extremely simple. The user enters a number and sees the quote related to that number. The quotes are stored in a `quotes` tuple. This is the data that normally exists in a database, file, and so on, and only the model has direct access to it.

Let's consider this example for the `quotes` tuple:

```
quotes = (
    "A man is not complete until he is married. Then he is finished.",
    "As I said before, I never repeat myself.",
```

```
        "Behind a successful man is an exhausted woman.",
        "Black holes really suck...",
        "Facts are stubborn things.",
    )
```

The model is minimalistic; it only has a `get_quote()` method that returns the quote (string) of the `quotes` tuple based on its index, n. The model class is as follows:

```
class QuoteModel:
    def get_quote(self, n):
        try:
            value = quotes[n]
        except IndexError as err:
            value = "Not found!"
        return value
```

The view has three methods: `show()`, which is used to print a quote (or the `Not found!` message) on the screen; `error()`, which is used to print an error message on the screen; and `select_quote()`, which reads the user's selection. This can be seen in the following code:

```
class QuoteTerminalView:
    def show(self, quote):
        print(f'And the quote is: "{quote}"')

    def error(self, msg):
        print(f"Error: {msg}")

    def select_quote(self):
        return input("Which quote number would you like to see? ")
```

The controller does the coordination. The `__init__()` method initializes the model and view. The `run()` method validates the quoted index given by the user, gets the quote from the model, and passes it back to the view to be displayed, as shown in the following code:

```
class QuoteTerminalController:
    def __init__(self):
        self.model = QuoteModel()
        self.view = QuoteTerminalView()

    def run(self):
        valid_input = False
        while not valid_input:
            try:
                n = self.view.select_quote()
                n = int(n)
```

```
            valid_input = True
        except ValueError as err:
            self.view.error(f"Incorrect index '{n}'")
    quote = self.model.get_quote(n)
    self.view.show(quote)
```

Finally, the `main()` function initializes and fires the controller, as shown in the following code:

```
def main():
    controller = QuoteTerminalController()
    while True:
        controller.run()
```

Here is a recap of our example (the full code is in `ch06/mvc.py` file):

1. We start by defining a variable for the list of quotes.

2. We define the model class, `QuoteModel`.

3. We define the view class, `QuoteTerminalView`.

4. We define the controller class, `QuoteTerminalController`.

5. Finally, we add the `main()` function to test the different classes, followed by the usual trick to call it.

A sample execution of the `python ch06/mvc.py` command shows how the program prints quotes to the user:

```
Which quote number would you like to see? 3
And the quote is: "Black holes really suck..."
Which quote number would you like to see? 2
And the quote is: "Behind a successful man is an exhausted woman."
Which quote number would you like to see? 6
And the quote is: "Not found!"
Which quote number would you like to see? 4
And the quote is: "Facts are stubborn things."
Which quote number would you like to see? 3
And the quote is: "Black holes really suck..."
Which quote number would you like to see? 1
And the quote is: "As I said before, I never repeat myself."
```

The Microservices pattern

Traditionally, developers working on building a server-side application have been using a single code base and implementing all or most functionalities right there, using common development practices such as functions and classes, and design patterns such as the ones we have covered in this book so far.

However, with the evolution of the IT industry, economic factors, and pressure for fast times to market and returns on investment, there is a constant need to improve the practices of engineering teams and ensure more reactivity and scalability with servers, service delivery, and operations. We need to learn about other useful patterns, not only object-oriented programming ones.

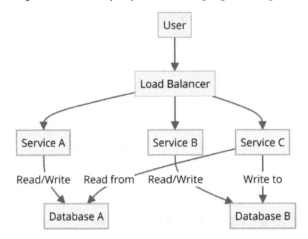

Figure 6.2 – The Microservices pattern

One of the main additions to the catalog of patterns for engineers in recent years has been the **Microservice Architecture** pattern or **Microservices**. The idea is that we can build an application as a set of loosely coupled, collaborating services. In this architectural style, an application might consist of services such as the order management service, the customer management service, and so on. These services are loosely coupled, independently deployable, and communicate via well-defined APIs.

Real-world examples

We can cite several examples, such as the following:

- **Netflix**: One of the pioneers in adopting microservices to handle millions of content streams simultaneously

- **Uber**: The company uses microservices to handle different aspects such as billing, notifications, and ride tracking

- **Amazon**: They transitioned from a monolithic architecture to microservices to support their ever-growing scale

Use cases for the Microservices pattern

We can think of several use cases where Microservices offer a clever answer. We can use a Microservices architecture-based design every time we are building an application that has at least one of the following characteristics:

- There is a requirement to support different clients, including desktop and mobile

- There is an API for third parties to consume

- We must communicate with other applications using messaging

- We serve requests by accessing a database, communicating with other systems, and returning the right type of response (JSON, XML, HTML, or even PDF)

- There are logical components corresponding to different functional areas of the application

Implementing the microservices pattern – a payment service using gRPC

Let's briefly talk about software installation and application deployment in the Microservices world. Switching from deploying a single application to deploying many small services means that the number of things that need to be handled increases exponentially. While you might have been fine with a single application server and a few runtime dependencies, when moving to Microservices, the number of dependencies will increase drastically. For example, one service could benefit from the relational database while the other would need **ElasticSearch**. You may need a service that uses **MySQL** and another one that uses the **Redis** server. So, using the Microservices approach also means you will need to use **containers**.

Thanks to Docker, things have become easier, since we can run those services as containers. The idea is that your application server, dependencies and runtime libraries, compiled code, configurations, and so on, are inside those containers. Then, all you must do is run services packed as containers and make sure that they can communicate with each other.

You can implement the Microservices pattern, for a web app or an API, by directly using Django, Flask, or FastAPI. However, to quickly show a working example, we are going to use gRPC, a high-performance universal RPC framework that uses **Protocol Buffers (protobuf)** as its interface description language, making it an ideal candidate for microservices communication due to its efficiency and cross-language support.

Imagine a scenario where your application architecture includes a microservice dedicated to handling payment processing. This microservice (let's call it `PaymentService`), is responsible for processing payments and interacts with other services such as `OrderService` and `AccountService`. We are going to focus on the implementation of such a service using gRPC.

First, we define the service and its methods using protobuf, in the ch06/microservices/grpc/ payment.proto file. This includes specifying request and response message formats:

```
syntax = "proto3";

package payment;

// The payment service definition.
service PaymentService {
  // Processes a payment
  rpc ProcessPayment (PaymentRequest) returns (PaymentResponse) {}
}

// The request message containing payment details.
message PaymentRequest {
  string order_id = 1;
  double amount = 2;
  string currency = 3;
  string user_id = 4;
}

// The response message containing the result of the payment process.
message PaymentResponse {
  string payment_id = 1;
  string status = 2; // e.g., "SUCCESS", "FAILED"
}
```

Then, you must compile the payment.proto file into Python code using the protobuf compiler (protoc). For that, you need to use a specific command line that invokes protoc with the appropriate plugins and options for Python.

Here is the general form of the command line for compiling .proto files for use with gRPC in Python:

```
python -m grpc_tools.protoc -I<PROTO_DIR> --python_out=<OUTPUT_DIR>
--grpc_python_out=<OUTPUT_DIR> <PROTO_FILES>
```

In this case, we make sure we change the directory to be under the right path (for example, by doing cd ch06/microservices/grpc), and then we run the following command:

```
python -m grpc_tools.protoc -I. --python_out=. --grpc_python_out=.
payment.proto
```

This will generate two files in the current directory: payment_pb2.py and payment_pb2_grpc. py. Those files are not to be manually edited.

Next, we provide, in a `payment_service.py` file, the service logic for the payment processing, extending what has been provided in the generated `.py` files. In the module, we define the `PaymentServiceImpl` class, inheriting from the `payment_pb2_grpc.PaymentServiceServicer` class, and we override the `ProcessPayment()` method that will do what is needed to process the payment (e.g., calling external APIs, doing database updates, etc.) Note that here, we have a simplified example, but you would have more complex logic. The code is as follows:

```
from concurrent.futures import ThreadPoolExecutor
import grpc
import payment_pb2
import payment_pb2_grpc


class PaymentServiceImpl(payment_pb2_grpc.PaymentServiceServicer):
    def ProcessPayment(self, request, context):
        return payment_pb2.PaymentResponse(payment_id="12345",
status="SUCCESS")
```

Then, we have the `main()` function, with the code needed to start the processing service, created by calling `grpc.server(ThreadPoolExecutor(max_workers=10))`. The code of the function is as follows:

```
def main():
    print("Payment Processing Service ready!")
    server = grpc.server(ThreadPoolExecutor(max_workers=10))
    payment_pb2_grpc.add_PaymentServiceServicer_to_
server(PaymentServiceImpl(), server)
    server.add_insecure_port("[::]:50051")
    server.start()
    server.wait_for_termination()
```

With that, the service is done and ready to be tested. We need a client to be able to test it. We can write a test client with code that calls the service using gRPC, with the following code (in the `ch06/microservices/grpc/client.py` file):

```
import grpc
import payment_pb2
import payment_pb2_grpc


with grpc.insecure_channel("localhost:50051") as chan:
    stub = payment_pb2_grpc.PaymentServiceStub(chan)
    resp = stub.ProcessPayment(
        payment_pb2.PaymentRequest(
            order_id="order123",
```

```
            amount=99.99,
            currency="USD",
            user_id="user456",
        )
    )
    print("Payment Service responded.")
    print(f"Response status: {resp.status}")
```

To start the service (in the `ch06/microservices/grpc/payment_service.py` file), you can run the following command:

```
python ch06/microservices/grpc/payment_service.py
```

You will get the following output, showing that the service has started as expected:

```
Payment Processing Service ready!
```

Now, open another terminal to run the client (in the `ch06/microservices/grpc/client.py` file):

```
python ch06/microservices/grpc/client.py
```

In the terminal where you have run the client code, you should get the following output:

```
Payment Service responded.
Response status: SUCCESS
```

This output is what is expected.

Note that while gRPC is a powerful choice for Microservices communication, other approaches such as **REST over HTTP** can also be used, especially when human readability or web integration is a priority. However, gRPC provides advantages in terms of performance and support for streaming requests and responses, and it was interesting to introduce it with this example.

Implementing the microservices pattern – an LLM service using Lanarky

Lanarky is a web framework that builds upon the FastAPI framework, to provide batteries for building Microservices that use **large language models** (**LLMs**).

We will follow the *Getting started* instructions from the website (`https://lanarky.ajndkr.com`) to showcase a microservice backed by Lanarky. To be able to test the example, you need to set the `OPENAI_API_KEY` environment variable to use OpenAI. Visit `https://openai.com` and follow the instructions to get your API key.

The LLM service code starts by importing the modules we need:

```
import os

import uvicorn
from lanarky import Lanarky
from lanarky.adapters.openai.resources import ChatCompletionResource
from lanarky.adapters.openai.routing import OpenAIAPIRouter
```

Before starting the actual application code, you need to pass the OpenAI API key, which is used by Lanarky's code via the os.environ object. For example, pass the value of the secret key via this line:

```
os.environ["OPENAI_API_KEY"] = "Your OpenAI API key here"
```

> **Security practice**
>
> It is recommended that you pass secret keys to the code, by setting an environment variable in your shell.

Then, we create an app object, an instance of the Lanarky class, and the router object that will be used for the definition of the service's routes, as is conventional with FastAPI. This router is an instance of the OpenAPIRouter class provided by the Lanarky framework:

```
app = Lanarky()
router = OpenAIAPIRouter()
```

Next, we provide a chat() function for the /chat route, when there is a POST request, as follows:

```
@router.post("/chat")
def chat(stream: bool = True) -> ChatCompletionResource:
    system = "Here is your assistant"
    return ChatCompletionResource(stream=stream, system=system)
```

Finally, we associate the router to the FastAPI application (standard FastAPI convention) and we run the FastAPI application (our service) using uvicorn.run(), as follows:

```
if __name__ == "__main__":
    app.include_router(router)
    uvicorn.run(app)
```

To finalize this demonstration implementation, we can write client code to interact with the service. The code for that part is as follows:

```
import click
import sys
```

```
from lanarky.clients import StreamingClient


args = sys.argv[1:]
if len(args) == 1:
    message = args[0]

    client = StreamingClient()
    for event in client.stream_response(
        "POST",
        "/chat",
        params={"stream": "false"},
        json={"messages": [dict(role="user", content=message)]},
    ):
        print(f"{event.event}: {event.data}")
else:
    print("You need to pass a message!")
```

To test the example, similarly to the previous one (where we tested a gRPC-based microservice), open a terminal, and run the LLM service code (in the ch06/microservices/lanarky/llm_service.py file) using the following command:

```
python ch06/microservices/lanarky/llm_service.py
```

You should get an output like the following:

```
INFO:    Started server process [18617]
INFO:    Waiting for application startup.
INFO:    Application startup complete.
INFO:    Uvicorn running on http://127.0.0.1:8000 (Press CTRL+C to
quit)
```

Then, open a second terminal to run the client program, using the following command:

```
python ch06/microservices/lanarky/client.py "Hello"
```

You should get the following output:

```
completion: Hello! How can I assist you today?
```

Now, you can continue sending messages via the client program, and wait for the service to come back with the completion, as you would do via the ChatGPT interface.

For example, see the following code:

```
python ch06/microservices/lanarky/client.py "What is the capital of
Switzerland?"
completion: The capital of Switzerland is Bern.
```

The Serverless pattern

The Serverless pattern abstracts server management, allowing developers to focus solely on code. Cloud providers handle the scaling and execution based on event triggers, such as HTTP requests, file uploads, or database modifications.

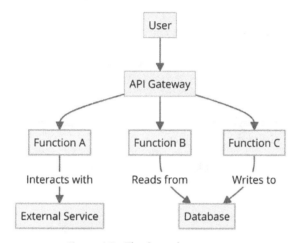

Figure 6.3 – The Serverless pattern

The Serverless pattern is particularly useful for Microservices, APIs, and event-driven architectures.

Real-world examples

There are several examples we can think of for the Serverless pattern. Here are some of them:

- **Automated data backups**: Serverless functions can be scheduled to automatically back up important data to cloud storage

- **Image processing**: Whenever a user uploads an image, a serverless function can automatically resize, compress, or apply filters to the image

- **PDF generation for E-commerce receipts**: After a purchase is made, a serverless function generates a PDF receipt and emails it to the customer

Use cases for the Serverless pattern

There are two types of use cases the Serverless pattern can be used for.

First, Serverless is useful for handling event-driven architectures where specific functions need to be executed in response to events, such as doing image processing (cropping, resizing) or dynamic PDF generation.

The second type of architecture where Serverless can be used is **Microservices**. Each microservice can be a serverless function, making it easier to manage and scale.

Since we have already discussed the Microservices pattern in the previous section, we are going to focus on how to implement the first use case.

Implementing the Serverless pattern

Let's see a simple example using AWS Lambda to create a function that squares a number. AWS Lambda is Amazon's serverless **compute** service, which runs code in response to triggers such as changes in data, shifts in system state, or actions by users.

There is no need to add more complexity since there's already enough to get right with the Serverless architecture itself and AWS Lambda's deployment details.

First, we need to write the Python code for the function. We create a `lambda_handler()` function, which takes two parameters, `event` and `context`. In our case, the input number is accessed as a value of the "number" key in the event dictionary. We take the square of that value and we return a a string containing the expected result. The code is as follows:

```
import json


def lambda_handler(event, context):
    number = event["number"]
    squared = number * number
    return f"The square of {number} is {squared}."
```

Once we have the Python function, we need to deploy it so that it can be invoked as an AWS Lambda function. For our learning, instead of going through the procedure of deploying to AWS infrastructure, we can use a method that consists of testing things locally. This is what the `LocalStack` Python package allows us to do. Once it is installed, from your environment, you can start LocalStack inside a Docker container by running the available executable in your Python environment, using the command:

```
localstack start -d
```

Then, we compress our Python code file (ch06/lambda_function_square.py) to a ZIP file, for example, by using the ZIP program as follows:

```
zip lambda.zip lambda_function_square.py
```

The other tool we are going to use here is the awslocal tool (a Python module we need to install). Once installed, we can use this program to deploy the Lambda function into the "local stack" AWS infrastructure. This is done, in our case, using the following command:

```
awslocal lambda create-function \
    --function-name lambda_function_square \
    --runtime python3.11 \
    --zip-file fileb://lambda.zip \
    --handler lambda_function_square.lambda_handler \
    --role arn:aws:iam::000000000000:role/lambda-role
```

> **Adapt to your Python version**
>
> At the time of writing, this was tested with Python 3.11. You must adapt this command to your Python version.

You can test the Lambda function, providing an input using the payload.json file, using the command:

```
awslocal lambda invoke --function-name lambda_function_square \
    --payload file://payload.json output.txt
```

You can then check the result by looking into the output.txt file's content. You should see the text:

```
The square of 6 is 36.
```

Okay, this was a preliminary test, but we can go further. We can create a URL for the Lambda function. Again, thanks to awslocal, running the following command:

```
awslocal lambda create-function-url-config \
    --function-name lambda_function_square \
    --auth-type NONE
```

This will generate a URL that can be used to invoke the Lambda function. The URL will be in the http://<XXXXXXXX>.lambda-url.us-east-1.localhost.localstack.cloud:4566 format.

Now, for example, we can trigger the Lambda function URL using cUrl:

```
curl -X POST \
    'http://iu4s187onr1oabg50dbvm77bk6r5sunk.lambda-url.us-east-1.
localhost.localstack.cloud:4566/' \
```

```
    -H 'Content-Type: application/json' \
    -d '{"number": 6}'
```

For up-to-date and detailed guides related to AWS Lambda, consult the documentation at `https://docs.aws.amazon.com/lambda/`.

This was a minimal example. Another example of a serverless application could be a function that generates PDF receipts for a business. This would allow the business to not worry about server management and only pay for the computing time that is consumed.

The Event Sourcing pattern

The Event Sourcing pattern stores state changes as a sequence of events, allowing the reconstruction of past states and providing an audit trail. This pattern is particularly useful in systems where the state is complex and the business rules for transitions are complex.

As we will see in implementation examples later, the Event Sourcing pattern emphasizes the importance of capturing all changes to an application state as a sequence of events. An outcome of this is that the application state can be reconstructed at any point in time by replaying these events.

Real-world examples

There are several real-world examples in the software category:

- **Audit trails**: Keeping a record of all changes made to a database for compliance
- **Collaborative editing**: Allowing multiple users to edit a document simultaneously
- **Undo/redo features**: Providing the ability to undo or redo actions in an application

Use cases for the Event Sourcing pattern

There are several use cases for the Event Sourcing pattern. Let's consider the following three:

- **Financial transactions**: Event Sourcing can be used to record every change to an account's balance as a chronological series of immutable events. This method ensures that every deposit, withdrawal, or transfer is captured as a distinct event. This way, we can provide a transparent, auditable, and secure ledger of all financial activities.
- **Inventory management**: Within inventory management contexts, Event Sourcing helps in tracking each item's life cycle by logging all changes as events. This enables businesses to maintain accurate and up-to-date records of stock levels, identify patterns in item usage or sales, and predict future inventory needs. It also facilitates tracing the history of any item, aiding in recall processes or quality assurance investigations.

- **Customer behavior tracking**: Event Sourcing plays a critical role in capturing and storing every interaction a customer has with a platform, from browsing history and cart modifications to purchases and returns. This wealth of data, structured as a series of events, becomes a valuable resource for analyzing customer behavior, personalizing marketing strategies, enhancing user experience, and improving product recommendations.

Let's now see how we can implement this pattern.

Implementing the event sourcing pattern – the manual way

Let's start with some definitions. The components of the Event Sourcing pattern implementation are as follows:

- **Event**: A representation of a state change, typically containing the type of event and the data associated with that event. Once an event is created and applied, it cannot be changed.

- **Aggregate**: An object (or group of objects) that represents a single unit of business logic or data. It keeps track of things, and every time something changes (an event), it makes a record of it.

- **Event store**: A collection of all the events that have occurred.

By handling state changes through events, the business logic becomes more flexible and easier to extend. For example, adding new types of events or modifying the handling of existing events can be done with minimal impact on the rest of the system.

In this first example, for the bank account use case, we will see how to implement the event sourcing pattern in a manual way. In such an implementation, you would typically define your event classes and manually write the logic to apply these events to your aggregates. Let's see that.

We start by defining an `Account` class representing a bank account with a balance and a list of events attached to it, for the operations on the account. This class acts as the aggregate. Its `events` attribute represents the event store. Here, an event will be represented by a dictionary containing the type of operation ("deposited" or "withdrawn") and the amount value.

We then add the `apply_event()` method taking an event as the input. Depending on `event["type"]`, we increment or decrement the account balance by the event's amount, and we add the event to the `events` list, effectively storing the event:

```python
class Account:
    def __init__(self):
        self.balance = 0
        self.events = []

    def apply_event(self, event):
        if event["type"] == "deposited":
            self.balance += event["amount"]
```

```
        elif event["type"] == "withdrawn":
            self.balance -= event["amount"]
        self.events.append(event)
```

Then, we add a `deposit()` method and a `withdraw()` method, which both call the `apply_event()` method, as follows:

```
def deposit(self, amount):
    event = {"type": "deposited", "amount": amount}
    self.apply_event(event)

def withdraw(self, amount):
    event = {"type": "withdrawn", "amount": amount}
    self.apply_event(event)
```

Finally, we add the `main()` function, as follows:

```
def main():
    account = Account()
    account.deposit(100)
    account.deposit(50)
    account.withdraw(30)
    account.deposit(30)

    for evt in account.events:
        print(evt)
    print(f"Balance: {account.balance}")
```

Running the code, using the `python ch06/ event_sourcing/bankaccount.py` command, gives the following output:

```
{'type': 'deposited', 'amount': 100}
{'type': 'deposited', 'amount': 50}
{'type': 'withdrawn', 'amount': 30}
{'type': 'deposited', 'amount': 30}
Balance: 150
```

This example provided a first understanding of Event Sourcing through a simple, manual implementation. For more complex systems, frameworks and libraries designed for Event Sourcing can help manage some of this complexity, providing utilities for event storage, querying, and processing. We will test such a library next.

Implementing the Event Sourcing pattern – using a library

In this second example, we will use the `eventsourcing` library to implement the Event Sourcing pattern. Let's consider an inventory management system where we track the quantity of items.

We start by importing what we need, as follows:

```
from eventsourcing.domain import Aggregate, event
from eventsourcing.application import Application
```

Then, we define the class for the aggregate object, `InventoryItem`, by inheriting from the `Aggregate` class. The class has an `increase_quantity()` and a `decrease_quantity` method, each decorated with the `@event` decorator. The code for this class is as follows:

```
class InventoryItem(Aggregate):
    @event("ItemCreated")
    def __init__(self, name, quantity=0):
        self.name = name
        self.quantity = quantity

    @event("QuantityIncreased")
    def increase_quantity(self, amount):
        self.quantity += amount

    @event("QuantityDecreased")
    def decrease_quantity(self, amount):
        self.quantity -= amount
```

Next, we create our inventory application's class, `InventoryApp`, inheriting from the `eventsourcing` library's `Application` class. The first method handles the creation of an item, taking an instance of the `InventoryItem` class (`item`) and calling the `save()` method on the `InventoryApp` object using the item. But what exactly does the `save()` method do? It collects pending events from given aggregates and puts them in the application's event store. The definition of the class starts as follows:

```
class InventoryApp(Application):
    def create_item(self, name, quantity):
        item = InventoryItem(name, quantity)
        self.save(item)
        return item.id
```

Next, similarly to what we did in the previous example, we add an `increase_item_quantity()` method, which handles the increase of the item's quantity (for the aggregate object) and then saves the aggregate object on the application, followed by the corresponding `decrease_item_quantity()` method, for the decreasing action, as follows:

```python
def increase_item_quantity(self, item_id, amount):
    item = self.repository.get(item_id)
    item.increase_quantity(amount)
    self.save(item)

def decrease_item_quantity(self, item_id, amount):
    item = self.repository.get(item_id)
    item.decrease_quantity(amount)
    self.save(item)
```

Finally, we add the `main()` function, with some code to test our design, as follows:

```python
def main():
    app = InventoryApp()

    # Create a new item
    item_id = app.create_item("Laptop", 10)
    # Increase quantity
    app.increase_item_quantity(item_id, 5)
    # Decrease quantity
    app.decrease_item_quantity(item_id, 3)

    notifs = app.notification_log.select(start=1, limit=5)
    notifs = [notif.state for notif in notifs]
    for notif in notifs:
        print(notif.decode())
```

Running the code, using the `python ch06/ event_sourcing/inventory.py` command, gives the following output:

```
{"timestamp":{"_type_":"datetime_iso","_data_":"2024-
03-18T08:05:10.583875+00:00"},"originator_topic":"__
main__:InventoryItem","name":"Laptop","quantity":10}
{"timestamp":{"_type_":"datetime_iso","_data_":"2024-03-
18T08:05:10.584818+00:00"},"amount":5}
{"timestamp":{"_type_":"datetime_iso","_data_":"2024-03-
18T08:05:10.585128+00:00"},"amount":3}
```

Nice work! This example and the previous one helped introduce the way to design an event-sourced application. For an ambitious project, we can leverage the `eventsourcing` library, which makes it easier to implement this type of application.

Other architectural design patterns

You may encounter documentation about other architectural design patterns. Here are three other patterns:

- **Event-Driven Architecture (EDA)**: This pattern emphasizes the production, detection, consumption of, and reaction to events. EDA is highly adaptable and scalable, making it suitable for environments where systems need to react to significant events in real time.

- **Command Query Responsibility Segregation (CQRS)**: This pattern separates the models for reading and writing data, allowing for more scalable and maintainable architectures, especially when there are clear distinctions between operations that mutate data and those that only read data.

- **Clean Architecture**: This pattern proposes a way to organize code such that it encapsulates the business logic but keeps it separate from the interfaces through which the application is exposed to users or other systems. It emphasizes the use of dependency inversion to drive the decoupling of software components.

Summary

In this chapter, we explored several foundational architectural design patterns that are pivotal in modern software development, each useful for different requirements and solving unique challenges.

We first covered the MVC pattern, which promotes the separation of concerns by dividing the application into three interconnected components. This separation allows for more manageable, scalable, and testable code by isolating the UI, the data, and the logic that connects the two.

Then, we looked at the Microservices pattern, which takes a different approach by structuring an application as a collection of small, independent services, each responsible for a specific business function. This pattern enhances scalability, flexibility, and ease of deployment, making it an ideal choice for complex, evolving applications that need to rapidly adapt to changing business requirements.

Next, we looked at the Serverless pattern, which shifts the focus from server management to pure business logic by leveraging cloud services to execute code snippets in response to events. This pattern offers significant cost savings, scalability, and productivity benefits by abstracting the underlying infrastructure, allowing developers to concentrate on writing code that adds direct value.

Afterward, we went over the Event Sourcing pattern, which offers another way to handle data changes in an application by storing each change as a sequence of events. This not only provides a robust audit trail and enables complex business functionalities but also allows the system to reconstruct past states, offering invaluable insights into the data life cycle and changes over time.

Lastly, we touched upon other architectural design patterns, such as CQRS and Clean Architecture. Each offers unique advantages and addresses different aspects of software design and architecture. Even if we could not dive deep into these patterns, they complement the developer's toolkit for building well-structured and maintainable systems.

In the next chapter, we will discuss concurrency and asynchronous patterns and techniques to help our program manage multiple operations simultaneously or move on to other tasks while waiting for operations to complete.

7
Concurrency and Asynchronous Patterns

In the previous chapter, we covered architectural design patterns: patterns that help with solving some unique challenges that come with complex projects. Next, we need to discuss concurrency and asynchronous patterns, another important category in our solutions catalog.

Concurrency allows your program to manage multiple operations simultaneously, leveraging the full power of modern processors. It's akin to a chef preparing multiple dishes in parallel, each step orchestrated so that all dishes are ready at the same time. Asynchronous programming, on the other hand, lets your application move on to other tasks while waiting for operations to complete, such as sending a food order to the kitchen and serving other customers until the order is ready.

In this chapter, we're going to cover the following main topics:

- The Thread Pool pattern
- The Worker Model pattern
- The Future and Promise pattern
- The Observer pattern in reactive programming
- Other concurrency and asynchronous patterns

Technical requirements

See the requirements presented in *Chapter 1*. The additional technical requirements for the code discussed in this chapter are the following:

- Faker, using `pip install faker`
- ReactiveX, using `pip install reactivex`

The Thread Pool pattern

First, it's important to understand what a thread is. In computing, a thread is the smallest unit of processing that can be scheduled by an operating system.

Threads are like tracks of execution that can run on a computer at the same time, which enables many activities to be done simultaneously and thus improve performance. They are particularly important in applications that need multitasking, such as serving multiple web requests or carrying out multiple computations.

Now, onto the Thread Pool pattern itself. Imagine you have many tasks to complete but starting each task (which means in this case, creating a thread) can be expensive in terms of resources and time. It's like hiring a new employee every time you have a job to do and then letting them go when the job is done. This process can be inefficient and costly. By maintaining a collection, or a pool, of worker threads that can be created for once and then reused upon several jobs, the Thread Pool pattern helps reduce this inefficiency. When one thread finishes a task, it does not terminate but goes back to the pool, awaiting another task that it can be used again for.

> **What are worker threads?**
>
> A worker thread is a thread of execution of a particular task or set of tasks. Worker threads are used to offload processing tasks from the main thread, helping to keep applications responsive by performing time-consuming or resource-intensive tasks asynchronously.

In addition to faster application performance, there are two benefits:

- **Reduced overhead**: By reusing threads, the application avoids the overhead of creating and destroying threads for each task
- **Better resource management**: The thread pool limits the number of threads, preventing resource exhaustion that could occur if too many threads were created

Real-world examples

In real life, imagine a small restaurant with a limited number of chefs (threads) who cook meals (tasks) for customers. The restaurant can only accommodate a certain number of chefs working at once due to kitchen space (system resources). When a new order comes in, if all chefs are busy, the order waits in a queue until there is an available chef. This way, the restaurant efficiently manages the flow of orders with its available chefs, ensuring all are utilized effectively without overwhelming the kitchen or needing to hire more staff for each new order.

There are also many examples in software:

- Web servers often use thread pools to handle incoming client requests. This allows them to serve multiple clients simultaneously without the overhead of creating a new thread for each request.

- Databases use thread pools to manage connections, ensuring that a pool of connections is always available for incoming queries.

- Task schedulers use thread pools to execute scheduled tasks such as *cron* jobs, backups, or updates.

Use cases for the Thread Pool pattern

There are three use cases where the Thread Pool pattern helps:

- **Batch processing**: When you have many tasks that can be performed in parallel, a thread pool can distribute them among its worker threads

- **Load balancing**: Thread pools can be used to distribute workload evenly among worker threads, ensuring that no single thread takes on too much work

- **Resource optimization**: By reusing threads, the thread pool minimizes system resource usage, such as memory and CPU time

Implementing the Thread Pool pattern

First, let's stop to break down how a thread pool, for a given application, works:

1. When the application starts, the thread pool creates a certain number of worker threads. This is the initialization. This number of threads can be fixed or dynamically adjusted based on the application's needs.

2. Then, we have the task submission step. When there's a task to be done, it's submitted to the pool rather than directly creating a new thread. The task can be anything that needs to be executed, such as processing user input, handling network requests, or performing calculations.

3. The following step is the task execution. The pool assigns the task to one of the available worker threads. If all threads are busy, the task might wait in a queue until a thread becomes available.

4. Once a thread completes its task, it doesn't die. Instead, it returns to the pool, ready to be assigned a new task.

For our example, let's see some code where we create a thread pool with five worker threads to handle a set of tasks. We are going to use the `ThreadPoolExecutor` class from the `concurrent.futures` module.

We start by importing what we need for the example, as follows:

```
from concurrent.futures import ThreadPoolExecutor
import time
```

Then, we create a function to simulate the tasks, by simply using `time.sleep(1)` in this case:

```
def task(n):
    print(f"Executing task {n}")
    time.sleep(1)
    print(f"Task {n} completed")
```

Then, we use an instance of the `ThreadPoolExecutor` class, created with a maximum number of worker threads of 5, and we submit 10 tasks to the thread pool. So, the worker threads pick up these tasks and execute them. Once a worker thread completes a task, it picks up another from the queue. The code is as follows:

```
with ThreadPoolExecutor(max_workers=5) as executor:
    for i in range(10):
        executor.submit(task, i)
```

When running the example code, using the `ch07/thread_pool.py` Python command, you should get the following output:

```
Executing task 0
Executing task 1
Executing task 2
Executing task 3
Executing task 4
Task 0 completed
Task 4 completed
Task 3 completed
Task 1 completed
Executing task 6
Executing task 7
Executing task 8
Task 2 completed
Executing task 5
Executing task 9
Task 8 completed
Task 6 completed
Task 9 completed
Task 5 completed
Task 7 completed
```

We see that the tasks were completed in an order different from the order of submission. This shows that they were executed concurrently using the threads available in the thread pool.

The Worker Model pattern

The idea behind the Worker Model pattern is to divide a large task or many tasks into smaller, manageable units of work, called workers, that can be processed in parallel. This approach to concurrency and parallel processing not only accelerates processing time but also enhances the application's performance.

The workers could be threads within a single application (as we have just seen in the Thread Pool pattern), separate processes on the same machine, or even different machines in a distributed system.

The benefits of the Worker Model pattern are the following:

- **Scalability**: Easily scales with the addition of more workers, which can be particularly beneficial in distributed systems where tasks can be processed on multiple machines
- **Efficiency**: By distributing tasks across multiple workers, the system can make better use of available computing resources, processing tasks in parallel
- **Flexibility**: The Worker Model pattern can accommodate a range of processing strategies, from simple thread-based workers to complex distributed systems spanning multiple servers

Real-world examples

Consider a delivery service where packages (tasks) are delivered by a team of couriers (workers). Each courier picks up a package from the distribution center (task queue) and delivers it. The number of couriers can vary depending on demand; more couriers can be added during busy periods and reduced when it's quieter.

In big data processing, the Worker Model pattern is often employed where each worker is responsible for mapping or reducing a part of the data.

In systems such as RabbitMQ or Kafka, the Worker Model pattern is used to process messages from a queue concurrently.

We can also cite image processing services. Services that need to process multiple images simultaneously often use the Worker Model pattern to distribute the load among multiple workers.

Use cases for the Worker Model pattern

One use case for the Worker Model pattern is *data transformation*. When you have a large dataset that needs to be transformed, you can distribute the work among multiple workers.

Another one is *task parallelism*. In applications where different tasks are independent of each other, the Worker Model pattern can be very effective.

A third use case is *distributed computing*, where the Worker Model pattern can be extended to multiple machines, making it suitable for distributed computing environments.

Implementing the Worker Model pattern

Before discussing an implementation example, let's understand how the Worker Model pattern works. Three components are involved in the Worker Model pattern: workers, a task queue, and, optionally, a dispatcher:

- **The workers**: The primary actors in this model. Each worker can perform a piece of the task independently of the others. Depending on the implementation, a worker might process one task at a time or handle multiple tasks concurrently.

- **The task queue**: A central component where tasks are stored awaiting processing. Workers typically pull tasks from this queue, ensuring that tasks are distributed efficiently among them. The queue acts as a buffer, decoupling task submission from task processing.

- **The dispatcher**: In some implementations, a dispatcher component assigns tasks to workers based on availability, load, or priority. This can help optimize task distribution and resource utilization.

Let's now see an example where we execute a function in parallel.

We start by importing what we need for the example, as follows:

```
from multiprocessing import Process, Queue
import time
```

Then, we create a `worker()` function that we are going to run tasks with. It takes as a parameter the `task_queue` object that contains the tasks to execute. The code is as follows:

```
def worker(task_queue):
    while not task_queue.empty():
        task = task_queue.get()
        print(f"Worker {task} is processing")
        time.sleep(1)
        print(f"Worker {task} completed")
```

In the `main()` function, we start by creating a queue of tasks, an instance of `multiprocessing.Queue`. Then, we create 10 tasks and add them to the queue:

```
def main():
    task_queue = Queue()

    for i in range(10):
        task_queue.put(i)
```

Five worker processes are then created, using the `multiprocessing.Process` class, and started. Each worker picks up a task from the queue, to execute it, and then picks up another until the queue is empty. Then, we start each worker process (using `p.start()`) in a loop, which means that the associated task will get executed concurrently. After that, we create another loop where we use the process' `.join()` method so that the program waits for those processes to complete their work. That part of the code is as follows:

```
processes = [
    Process(target=worker, args=(task_queue,))
    for _ in range(5)
]

# Start the worker processes
for p in processes:
    p.start()

# Wait for all worker processes to finish
for p in processes:
    p.join()

print("All tasks completed.")
```

When running the example code, using the `ch07/worker_model.py` Python command, you should get the following output, where you can see that the 5 workers process tasks from the task queue in a concurrent way until all 10 tasks are completed:

```
Worker 0 is processing
Worker 1 is processing
Worker 2 is processing
Worker 3 is processing
Worker 4 is processing
Worker 0 completed
Worker 5 is processing
Worker 1 completed
Worker 6 is processing
Worker 2 completed
Worker 7 is processing
Worker 3 completed
Worker 8 is processing
Worker 4 completed
Worker 9 is processing
Worker 5 completed
Worker 6 completed
Worker 7 completed
```

```
Worker 8 completed
Worker 9 completed
All tasks completed.
```

This demonstrates our implementation of the Worker Model pattern. This pattern is particularly useful for scenarios where tasks are independent and can be processed in parallel.

The Future and Promise pattern

In the asynchronous programming paradigm, a Future represents a value that is not yet known but will be provided eventually. When a function initiates an asynchronous operation, instead of blocking until the operation completes and a result is available, it immediately returns a Future. This `Future` object acts as a placeholder for the actual result available later.

Futures are commonly used for I/O operations, network requests, and other time-consuming tasks that run asynchronously. They allow the program to continue executing other tasks rather than waiting for the operation to be completed. That property is referred to as *non-blocking*.

Once the Future is fulfilled, the result can be accessed through the Future, often via callbacks, polling, or blocking until the result is available.

A Promise is the writable, controlling counterpart to a Future. It represents the producer side of the asynchronous operation, which will eventually provide a result to its associated Future. When the operation completes, the Promise is fulfilled with a value or rejected with an error, which then resolves the Future.

Promises can be chained, allowing a sequence of asynchronous operations to be performed clearly and concisely.

By allowing a program to continue execution without waiting for asynchronous operations, applications become more responsive. Another benefit is *composability*: multiple asynchronous operations can be combined, sequenced, or executed in parallel in a clean and manageable way.

Real-world examples

Ordering a custom dining table from a carpenter provides a tangible example of the Future and Promise pattern. When you place the order, you receive an estimated completion date and design sketch (Future), representing the carpenter's promise to deliver the table. As the carpenter works, this promise moves toward fulfillment. The delivery of the completed table resolves the Future, marking the fulfillment of the carpenter's promise to you.

We can also find several examples in the digital realm, such as the following:

- **Online shopping order tracking**: When you place an order online, the website immediately provides you with an order confirmation and a tracking number (Future). As your order is processed, shipped, and delivered, status updates (Promise fulfillment) are reflected in real time on the tracking page, eventually resolving to a final delivery status.

- **Food delivery apps**: Upon ordering your meal through a food delivery app, you're given an estimated delivery time (Future). The app continuously updates the order status—from preparation through pickup and delivery (Promise being fulfilled)—until the food arrives at your door, at which point the Future is resolved with the completion of your order.

- **Customer support tickets**: When you submit a support ticket on a website, you immediately receive a ticket number and a message stating that someone will get back to you (Future). Behind the scenes, the support team addresses tickets based on priority or in the order they were received. Once your ticket is addressed, you receive a response, fulfilling the Promise made when you first submitted the ticket.

Use cases for the Future and Promise pattern

There are at least four use cases where the Future and Promise pattern is recommended:

1. **Data pipelines**: In data processing pipelines, data is often transformed through multiple stages before reaching its final form. By representing each stage with a Future, you can effectively manage the asynchronous flow of data. For example, the output of one stage can serve as the input for the next, but because each stage returns a Future, subsequent stages don't have to block while waiting for the previous ones to complete.

2. **Task scheduling**: Task scheduling systems, such as those in an operating system or a high-level application, can use Futures to represent tasks that are scheduled to run at a future time. When a task is scheduled, a Future is returned to represent the eventual completion of that task. This allows the system or the application to keep track of the task's state without blocking execution.

3. **Complex database queries or transactions**: Executing database queries asynchronously is crucial for maintaining application responsiveness, particularly in web applications where user experience is paramount. By using Futures to represent the outcome of database operations, applications can initiate a query and immediately return control to the user interface or the calling function. The Future will eventually resolve with the query result, allowing the application to update the UI or process the data without having frozen or become unresponsive while waiting for the database response.

4. **File I/O operations**: File I/O operations can significantly impact application performance, particularly if executed synchronously on the main thread. By applying the Future and Promise pattern, file I/O operations are offloaded to a background process, with a Future returned to represent the completion of the operation. This approach allows the application to continue running other tasks or responding to user interactions while the file is being read from or written to. Once the I/O operation completes, the Future resolves, and the application can process or display the file data.

In each of these use cases, the Future and Promise pattern facilitates asynchronous operation, allowing applications to remain responsive and efficient by not blocking the main thread with long-running tasks.

Implementing the Future and Promise pattern – using concurrent. futures

To understand how to implement the Future and Promise pattern, you must first understand the three steps of its mechanism. Let's break those down next:

1. **Initiation**: The initiation step involves starting an asynchronous operation using a function where, instead of waiting for the operation to complete, the function immediately returns a "Future" object. This object acts as a placeholder for the result that will be available later. Internally, the asynchronous function creates a "Promise" object. This object is responsible for handling the outcome of the asynchronous operation. The Promise is linked to the Future, meaning the state of the Promise (whether it's fulfilled or rejected) will directly affect the Future.

2. **Execution**: During the execution step, the operation proceeds independently of the main program flow. This allows the program to remain responsive and continue with other tasks. Once the asynchronous task completes, its result needs to be communicated back to the part of the program that initiated the operation. The outcome of the operation (be it a successful result or an error) is passed to the previously created Promise.

3. **Resolution**: If the operation is successful, the Promise is "fulfilled" with the result. If the operation fails, the Promise is "rejected" with an error. The fulfillment or rejection of the Promise resolves the Future. Using the result is often done through a callback or continuation function, which is a piece of code that specifies what to do with the result. The Future provides mechanisms (for example, methods or operators) to specify these callbacks, which will execute once the Future is resolved.

In our example, we use an instance of the `ThreadPoolExecutor` class to execute tasks asynchronously. The submit method returns a `Future` object that will eventually contain the result of the computation. We start by importing what we need, as follows:

```
from concurrent.futures import ThreadPoolExecutor, as_completed
```

Then, we define a function for the task to be executed:

```
def square(x):
    return x * x
```

We submit tasks and get `Future` objects, then we collect the completed Futures. The `as_completed` function allows us to iterate over completed `Future` objects and retrieve their results:

```
with ThreadPoolExecutor() as executor:
    future1 = executor.submit(square, 2)
    future2 = executor.submit(square, 3)
    future3 = executor.submit(square, 4)

    futures = [future1, future2, future3]

    for future in as_completed(futures):
        print(f"Result: {future.result()}")
```

When running the example, using the `ch07/future_and_promise/future.py` Python command, you should get the following output:

```
Result: 16
Result: 4
Result: 9
```

This demonstrates our implementation.

Implementing the Future and Promise pattern – using asyncio

Python's `asyncio` library provides another way to execute tasks using asynchronous programming. It is particularly useful for I/O-bound tasks. Let's see a second example using this technique.

What is asyncio?

The `asyncio` library provides support for asynchronous I/O, event loops, coroutines, and other concurrency-related tasks. So, using `asyncio`, developers can write code that efficiently handles I/O-bound operations.

Coroutines and async/await

A coroutine is a special kind of function that can pause and resume its execution at certain points, allowing other coroutines to run in the meantime. Coroutines are declared with the `async` keyword. Also, a coroutine can be awaited from other coroutines, using the `await` keyword.

We import the `asyncio` module, which contains everything we need:

```
import asyncio
```

Then, we create a function for the task of computing and returning the square of a number. We also want an I/O-bound operation, so we use `asyncio.sleep()`. Notice that in the `asyncio` style of programming, such a function is defined using the combined keywords `async def` – it is a coroutine. The `asyncio.sleep()` function itself is a coroutine, so we make sure to use the `await` keyword when calling it:

```python
async def square(x):
    # Simulate some IO-bound operation
    await asyncio.sleep(1)
    return x * x
```

Then, we move to creating our `main()` function. We use the `asyncio.ensure_future()` function to create the `Future` objects we want, passing it `square(x)`, with x being the number to square. We create three `Future` objects, `future1`, `future2`, and `future3`. Then, we use the `asyncio.gather()` coroutine to wait for our Futures to complete and gather the results. The code for the `main()` function is as follows:

```python
async def main():
    fut1 = asyncio.ensure_future(square(2))
    fut2 = asyncio.ensure_future(square(3))
    fut3 = asyncio.ensure_future(square(4))

    results = await asyncio.gather(fut1, fut2, fut3)

    for result in results:
        print(f"Result: {result}")
```

At the end of our code file, we have the usual `if __name__ == "__main__":` block. What is new here, since we are writing `asyncio`-based code, is that we need to run `asyncio`'s event loop, by calling `asyncio.run(main())`:

```python
if __name__ == "__main__":
    asyncio.run(main())
```

To test the example, run the `ch07/future_and_promise/async.py` Python command. You should get an output like the following:

```
Result: 4
Result: 9
Result: 16
```

The order of the results may vary, depending on who is running the program and when. In fact, it is not predictable. You may have noticed similar behavior in our previous examples. This is generally the case with concurrency or asynchronous code.

This simple example shows that `asyncio` is a suitable choice for the Future and Promise pattern when we need to efficiently handle I/O-bound tasks (in scenarios such as web scraping or API calls).

The Observer pattern in reactive programming

The Observer pattern (covered in *Chapter 5, Behavioral Design Patterns*) is useful for notifying an object or a group of objects when the state of a given object changes. This type of traditional Observer allows us to react to some object change events. It provides a nice solution for many cases, but in a situation where we must deal with many events, some depending on each other, the traditional way could lead to complicated, difficult-to-maintain code. That is where another paradigm called reactive programming gives us an interesting option. In simple terms, the concept of reactive programming is to react to many events (streams of events) while keeping our code clean.

Let's focus on ReactiveX (`http://reactivex.io`), which is a part of reactive programming. At the heart of ReactiveX is a concept known as an Observable. According to its official website, ReactiveX is about providing an API for asynchronous programming with what are called observable streams. This concept is added to the idea of the Observer, which we already discussed.

Imagine an Observable like a river that flows data or events down to an Observer. This Observable sends out items one after another. These items travel through a path made up of different steps or operations until they reach an Observer, who takes them in or consumes them.

Real-world examples

An airport's flight information display system is analogous to an Observable in reactive programming. Such a system continuously streams updates about flight statuses, including arrivals, departures, delays, and cancellations. This analogy illustrates how observers (travelers, airline staff, and airport services subscribed to receive updates) subscribe to an Observable (the flight display system) and react to a continuous stream of updates, allowing for dynamic responses to real-time information.

A spreadsheet application can also be seen as an example of reactive programming, based on its internal behavior. In virtually all spreadsheet applications, interactively changing any one cell in the sheet will result in immediately reevaluating all formulas that directly or indirectly depend on that cell and updating the display to reflect these reevaluations.

The ReactiveX idea is implemented in a variety of languages, including Java (RxJava), Python (RxPY), and JavaScript (RxJS). The Angular framework uses RxJS to implement the Observable pattern.

Use cases for the Observer pattern in reactive programming

One use case is the idea of a collection pipeline, discussed by Martin Fowler on his blog (`https://martinfowler.com/articles/collection-pipeline`).

> **Collection pipeline, described by Martin Fowler**
>
> Collection pipelines are a programming pattern where you organize some computation as a sequence of operations that compose by taking a collection as the output of one operation and feeding it into the next.

We can also use an Observable to do operations such as "map and reduce" or "groupby" on sequences of objects when processing data.

Finally, Observables can be created for diverse functions such as button events, requests, and Twitter feeds.

Implementing the Observer pattern in reactive programming

For this example, we decided to build a stream of a list of (fake) people's names (in the `ch07/observer_rx/people.txt`) text file, and an observable based on it.

> **Note**
>
> A first example text file containing fake names of people is provided (`ch07/observer_rx/people.txt`) as part of the book's example files. But a new one can be generated whenever needed using a helper script (`ch07/observer_rx/peoplelist.py`), which will be presented in a minute.

An example of such a list of names will look like this:

```
Peter Brown, Gabriel Hunt, Gary Martinez, Heather Fernandez, Juan
White, Alan George, Travis Davidson, David Adams, Christopher Morris,
Brittany Thomas, Brian Allen, Stefanie Lutz, Craig West, William
Phillips, Kirsten Michael, Daniel Brennan, Derrick West, Amy Vazquez,
Carol Howard, Taylor Abbott,
```

Back to our implementation. We start by importing what we need:

```
from pathlib import Path
import reactivex as rx
from reactivex import operators as ops
```

We define a function, `firstnames_from_db()`, which returns an Observable from the text file (reading the content of the file) containing the names, with transformations (as we have already seen) using `flat_map()`, `filter()`, and `map()` methods, and a new operation, `group_by()`, to emit items from another sequence—the first name found in the file, with its number of occurrence:

```
def firstnames_from_db(path: Path):
    file = path.open()

    # collect and push stored people firstnames
    return rx.from_iterable(file).pipe(
        ops.flat_map(
            lambda content: rx.from_iterable(
                content.split(", ")
            )
        ),
        ops.filter(lambda name: name != ""),
        ops.map(lambda name: name.split()[0]),
        ops.group_by(lambda firstname: firstname),
        ops.flat_map(
            lambda grp: grp.pipe(
                ops.count(),
                ops.map(lambda ct: (grp.key, ct)),
            )
        ),
    )
```

Then, in the `main()` function, we define an Observable that emits data every 5 seconds, merging its emission with what is returned from `firstnames_from_db(db_file)`, after setting `db_file` to the people names text file, as follows:

```
def main():
    db_path = Path(__file__).parent / Path("people.txt")

    # Emit data every 5 seconds
    rx.interval(5.0).pipe(
        ops.flat_map(lambda i: firstnames_from_db(db_path))
    ).subscribe(lambda val: print(str(val)))

    # Keep alive until user presses any key
    input("Starting... Press any key and ENTER, to quit\n")
```

Here is a recap of the example (complete code in the `ch07/observer_rx/rx_peoplelist.py` file):

1. We import the modules and classes we need.

2. We define a `firstnames_from_db()` function, which returns an Observable from the text file that is the source of the data. We collect and push the stored people's first names from that file.

3. Finally, in the main () function, we define an Observable that emits data every 5 seconds, merging its emission with what is returned from calling the firstnames_from_db() function.

To test the example, run the ch07/observer_rx/rx_peoplelist.py Python command. You should get an output like the following (only an extract is shown here):

```
Starting... Press any key and ENTER, to quit
('Peter', 1)
('Gabriel', 1)
('Gary', 1)
('Heather', 1)
('Juan', 1)
('Alan', 1)
('Travis', 1)
('David', 1)
('Christopher', 1)
('Brittany', 1)
('Brian', 1)
('Stefanie', 1)
('Craig', 1)
('William', 1)
('Kirsten', 1)
('Daniel', 1)
('Derrick', 1)
```

Once you press a key and press *Enter* on the keyboard, the emission is interrupted, and the program stops.

Handling new streams of data

Our test worked, but in a sense, it was static; the stream of data was limited to what is currently in the text file. What we need now is to generate several streams of data. The technique we can use to generate the type of fake data in the text file is based on a third-party module called Faker (https://pypi.org/project/Faker). The code that produces the data is provided to you, for free (in the ch07/observer_rx/peoplelist.py file), as follows:

```python
from faker import Faker
import sys


fake = Faker()


args = sys.argv[1:]
if len(args) == 1:
    output_filename = args[0]
```

```
        persons = []
        for _ in range(0, 20):
            p = {"firstname": fake.first_name(), "lastname": fake.last_
   name()}
            persons.append(p)

        persons = iter(persons)

        data = [f"{p['firstname']} {p['lastname']}" for p in persons]
        data = ", ".join(data) + ", "

        with open(output_filename, "a") as f:
            f.write(data)
    else:
        print("You need to pass the output filepath!")
```

Now, let's see what happens when we execute both programs (`ch07/observer_rx/peoplelist.py` and `ch07/observer_rx/rx_peoplelis.py`):

- From one command-line window or terminal, you can generate people's names, passing the right file path to the script; you would execute the following command: `python ch07/observer_rx/peoplelist.py ch07/observer_rx/people.txt`.

- From a second shell window, you can run the program that implements the Observable via the `python ch07/observer_rx/rx_peoplelist.py` command.

So, what is the output from both commands?

A new version of the `people.txt` file is created (with the random names in it, separated by a comma), to replace the existing file. And, each time you rerun that command (`python ch07/observer_rx/peoplelist.py`), a new set of names is added to the file.

The second command gives an output like the one you got with the first execution; the difference is that now it is not the same set of data that is emitted repeatedly. Now, new data can be generated in the source and emitted.

Other concurrency and asynchronous patterns

There are some other concurrency and asynchronous patterns developers may use. We can cite the following:

- **The Actor model**: A conceptual model to deal with concurrent computation. It defines some rules for how actor instances should behave: an actor can make local decisions, create more actors, send more messages, and determine how to respond to the next message received.

- **Coroutines**: General control structures where flow control is cooperatively passed between two different routines without returning. Coroutines facilitate asynchronous programming by allowing execution to be suspended and resumed. As we have seen in one of our examples, Python has coroutines built in (via the `asyncio` library).

- **Message passing**: Used in parallel computing, **object-oriented programming** (**OOP**), and **inter-process communication** (**IPC**), where software entities communicate and coordinate their actions by passing messages to each other.

- **Backpressure**: A mechanism to manage the flow of data through software systems and prevent overwhelming components. It allows systems to gracefully handle overload by signaling the producer to slow down until the consumer can catch up.

Each of these patterns has its use cases and trade-offs. It is interesting to know they exist, but we cannot discuss all the available patterns and techniques.

Summary

In this chapter, we discussed concurrency and asynchronous patterns, patterns useful for writing efficient, responsive software that can handle multiple tasks at once.

The Thread Pool pattern is a powerful tool in concurrent programming, offering a way to manage resources efficiently and improve application performance. It helps us improve application performance but also reduces overhead and better manages resources because the thread pool limits the number of threads.

While the Thread Pool pattern focuses on reusing a fixed number of threads to execute tasks, the Worker Model pattern is more about the dynamic distribution of tasks across potentially scalable and flexible worker entities. This pattern is particularly useful for scenarios where tasks are independent and can be processed in parallel.

The Future and Promise pattern facilitates asynchronous operation, allowing applications to remain responsive and efficient by not blocking the main thread with long-running tasks.

We also discussed the Observer pattern in reactive programming. The core idea of this pattern is to react to a stream of data and events, as with the streams of water we see in nature. We have lots of examples of this idea in the computing world. We have discussed an example of ReactiveX, which serves as an introduction for the reader to approach this programming paradigm and continue their own research via the ReactiveX official documentation.

Lastly, we touched upon the fact that there are other concurrency and asynchronous patterns. Each of these patterns has its use cases and trade-offs, but we cannot cover them all in a single book.

In the next chapter, we will discuss performance design patterns.

8

Performance Patterns

In the previous chapter, we covered concurrency and asynchronous patterns, useful for writing efficient software that can handle multiple tasks at once. Next, we are going to discuss specific performance patterns that help enhance the speed and resource utilization of applications.

Performance patterns address common bottlenecks and optimization challenges, providing developers with proven methodologies to improve execution time, reduce memory usage, and scale effectively.

In this chapter, we're going to cover the following main topics:

- The Cache-Aside pattern
- The Memoization pattern
- The Lazy Loading pattern

Technical requirements

See the requirements presented in *Chapter 1*. The additional technical requirements for the code discussed in this chapter are the following:

- Add the Faker module to your Python environment using the following command: `python -m pip install faker`
- Add the Redis module to your Python environment using the following command: `python -m pip install redis`
- Install the Redis server and run it using Docker: `docker run --name myredis -p 6379:6379 redis`

    If needed, follow the documentation at `https://redis.io/docs/latest/`

The Cache-Aside pattern

In situations where data is more frequently read than updated, applications use a cache to optimize repeated access to information stored in a database or data store. In some systems, that type of caching mechanism is built in and works automatically. When this is not the case, we must implement it in the application ourselves, using a caching strategy that is suitable for the particular use case.

One such strategy is called **Cache-Aside**, where, to improve performance, we store frequently accessed data in a cache, reducing the need to fetch data from the data store repeatedly.

Real-world examples

We can cite the following examples in the software realm:

- Memcached is commonly used as a cache server. It is a popular in-memory key-value store for small chunks of data from the results of database calls, API calls, or HTML page content.

- Redis is another server solution that is used for cache. Nowadays, it is my go-to server for caching or application in-memory storage use cases where it shines.

- Amazon's ElastiCache, according to the documentation site (`https://docs.aws.amazon. com/elasticache/`), is a web service that makes it easy to set up, manage, and scale a distributed in-memory data store or cache environment in the cloud.

Use cases for the cache-aside pattern

The cache-aside pattern is useful when we need to reduce the database load in our application. By caching frequently accessed data, fewer queries are sent to the database. It also helps improve application responsiveness, since cached data can be retrieved faster.

Note that this pattern works for data that doesn't change often and for data storage that doesn't depend on the consistency of a set of entries in the storage (multiple keys). For example, it might work for certain kinds of document stores or databases where keys are never updated and occasionally data entries are deleted but there is no strong requirement to continue to serve them for some time (until the cache is refreshed).

Implementing the cache-aside pattern

We can summarize the steps needed when implementing the Cache-Aside pattern, involving a database and a cache, as follows:

- **Case 1 – When we want to fetch a data item**: Return the item from the cache if found in it. If not found in the cache, read the data from the database. Put the item we got in the cache and return it.

- **Case 2 – When we want to update a data item**: Write the item in the database and remove the corresponding entry from the cache.

Let's try a simple implementation with a database of quotes from which the user can ask to retrieve some quotes via an application. Our focus here will be implementing the *Case 1* part.

Here are our choices for the additional software dependencies we need to install on the machine for this implementation:

- An SQLite database, since we can query an SQLite database using Python's standard module, `sqlite3`
- A Redis server and the `redis-py` Python module

We will use a script (in the `ch08/cache_aside/populate_db.py` file) to handle the creation of a database and a `quotes` table and add example data to it. For practical reasons, we also use the `Faker` module there to generate fake quotes that are used when populating the database.

Our code starts with the imports we need, followed by the creation of the Faker instance that we will use to generate fake quotes, as well as some constants or module-level variables:

```python
import sqlite3
from pathlib import Path
from random import randint

import redis
from faker import Faker

fake = Faker()

DB_PATH = Path(__file__).parent / Path("quotes.sqlite3")
cache = redis.StrictRedis(host="localhost", port=6379, decode_
responses=True)
```

Then, we write a function to take care of the database setup part, as follows:

```python
def setup_db():
    try:
        with sqlite3.connect(DB_PATH) as db:
            cursor = db.cursor()
            cursor.execute(
                """
                CREATE TABLE quotes(id INTEGER PRIMARY KEY, text TEXT)
                """
            )
```

```
            db.commit()
            print("Table 'quotes' created")
    except Exception as e:
        print(e)
```

Then, we define a central function that takes care of adding a set of new quotes based on a list of sentences or text snippets. Among different things, we associate a quote identifier to the quote, for the id column in the database table. To make things easier, we just pick a number randomly using quote_id = randint(1, 100). The add_quotes() function is defined as follows:

```
def add_quotes(quotes_list):
    added = []
    try:
        with sqlite3.connect(DB_PATH) as db:
            cursor = db.cursor()

            for quote_text in quotes_list:
                quote_id = randint(1, 100) # nosec
                quote = (quote_id, quote_text)

                cursor.execute(
                    """INSERT OR IGNORE INTO quotes(id, text)
VALUES(?, ?)""", quote
                )
                added.append(quote)

            db.commit()
    except Exception as e:
        print(e)

    return added
```

Next, we add a main() function, which in fact will have several parts; we want to use command-line argument parsing. Note the following:

- If we pass the init argument, we call the setup_db() function
- If we pass the update_all argument, we inject the quotes into the database and add them to the cache
- If we pass the update_db_only argument, we only inject the quotes into the database

The code of the `main()` function, called when running the Python script, is as follows:

```python
def main():
    msg = "Choose your mode! Enter 'init' or 'update_db_only' or
'update_all': "
    mode = input(msg)

    if mode.lower() == "init":
        setup_db()

    elif mode.lower() == "update_all":
        quotes_list = [fake.sentence() for _ in range(1, 11)]
        added = add_quotes(quotes_list)
        if added:
            print("New (fake) quotes added to the database:")
            for q in added:
                print(f"Added to DB: {q}")
                print("   - Also adding to the cache")
                cache.set(str(q[0]), q[1], ex=60)

    elif mode.lower() == "update_db_only":
        quotes_list = [fake.sentence() for _ in range(1, 11)]
        added = add_quotes(quotes_list)
        if added:
            print("New (fake) quotes added to the database ONLY:")
            for q in added:
                print(f"Added to DB: {q}")
```

That part is done. Now, we will create another module and script for the cache-aside-related operations themselves (in the `ch08/cache_aside/cache_aside.py` file).

We have a few imports needed here too, followed by constants:

```python
import sqlite3
from pathlib import Path

import redis

CACHE_KEY_PREFIX = "quote"
DB_PATH = Path(__file__).parent / Path("quotes.sqlite3")
cache = redis.StrictRedis(host="localhost", port=6379, decode_
responses=True)
```

Next, we define a `get_quote()` function to fetch a quote by its identifier. If we do not find the quote in the cache, we query the database to get it and we put the result in the cache before returning it. The function is defined as follows:

```python
def get_quote(quote_id: str) -> str:
    out = []
    quote = cache.get(f"{CACHE_KEY_PREFIX}.{quote_id}")

    if quote is None:
        # Get from the database
        query_fmt = "SELECT text FROM quotes WHERE id = {}"
        try:
            with sqlite3.connect(DB_PATH) as db:
                cursor = db.cursor()
                res = cursor.execute(query_fmt.format(quote_id)).fetchone()
                if not res:
                    return "There was no quote stored matching that id!"

                quote = res[0]
                out.append(f"Got '{quote}' FROM DB")
        except Exception as e:
            print(e)
            quote = ""

        # Add to the cache
        if quote:
            key = f"{CACHE_KEY_PREFIX}.{quote_id}"
            cache.set(key, quote, ex=60)
            out.append(f"Added TO CACHE, with key '{key}'")
    else:
        out.append(f"Got '{quote}' FROM CACHE")

    if out:
        return " - ".join(out)
    else:
        return ""
```

Finally, in the main part of the script, we ask for user input of a quote identifier, and we call `get_quote()` to fetch the quote. The code is as follows:

```python
def main():
    while True:
        quote_id = input("Enter the ID of the quote: ")
```

```
        if quote_id.isdigit():
            out = get_quote(quote_id)
            print(out)
        else:
            print("You must enter a number. Please retry.")
```

Now is the time to test our scripts, using the following steps.

First, by calling python ch08/cache_aside/populate_db.py, and choosing "init" for the mode option, we can see that a quotes.sqlite3 file is created (in the ch08/cache_aside/ folder), so we can conclude the database has been created and a quotes table created in it.

Then, we call python ch08/cache_aside/populate_db.py and pass the update_all mode; we get the following output:

```
Choose your mode! Enter 'init' or 'update_db_only' or 'update_all':
update_all
New (fake) quotes added to the database:
Added to DB: (62, 'Instead not here public.')
- Also adding to the cache
Added to DB: (26, 'Training degree crime serious beyond management
and.')
- Also adding to the cache
Added to DB: (25, 'Agree hour example cover game bed.')
- Also adding to the cache
Added to DB: (23, 'Dark team exactly really wind.')
- Also adding to the cache
Added to DB: (46, 'Only loss simple born remain.')
- Also adding to the cache
Added to DB: (13, 'Clearly statement mean growth executive mean.')
- Also adding to the cache
Added to DB: (88, 'West policy a human job structure bed.')
- Also adding to the cache
Added to DB: (25, 'Work maybe back play.')
- Also adding to the cache
Added to DB: (18, 'Here certain require consumer strategy.')
- Also adding to the cache
Added to DB: (48, 'Discover method many by hotel.')
- Also adding to the cache
```

We can also call python ch08/cache_aside/populate_db.py and choose the update_db_only mode. In that case, we get the following output:

```
Choose your mode! Enter 'init' or 'update_db_only' or 'update_all':
update_db_only
New (fake) quotes added to the database ONLY:
```

```
Added to DB: (73, 'Whose determine group what site.')
Added to DB: (77, 'Standard much career either will when chance.')
Added to DB: (5, 'Nature when event appear yeah.')
Added to DB: (81, 'By himself in treat.')
Added to DB: (88, 'Establish deal sometimes stage college everybody
close thank.')
Added to DB: (99, 'Room recently authority station relationship our
knowledge occur.')
Added to DB: (63, 'Price who a crime garden doctor eat.')
Added to DB: (43, 'Significant hot those think heart shake ago.')
Added to DB: (80, 'Understand and view happy.')
Added to DB: (54, 'Happen some family human involve.')
```

Next, we call the python ch08/cache_aside/cache_aside.py command, and we are asked for an input to try to fetch the matching quote. Here are the different outputs I got depending on the values I provided:

```
Enter the ID of the quote: 23
Got 'Dark team exactly really wind.' FROM DB - Added TO CACHE, with
key 'quote.23'
Enter the ID of the quote: 12
There was no quote stored matching that id!
Enter the ID of the quote: 43
Got 'Significant hot those think heart shake ago.' FROM DB - Added TO
CACHE, with key 'quote.43'
Enter the ID of the quote: 45
There was no quote stored matching that id!
Enter the ID of the quote: 77
Got 'Standard much career either will when chance.' FROM DB - Added TO
CACHE, with key 'quote.77'
```

So, each time I entered an identifier number that matched a quote stored only in the database (as shown by the previous output), the specific output showed that the data was obtained from the database first, before being returned from the cache (where it was immediately added).

We can see that things work as expected. The update part of the cache-aside implementation (to write the item in the database and remove the corresponding entry from the cache) is left to you to try. You could add an update_quote() function used to update a quote when you pass quote_id to it and call it using the right command line (such as python cache_aside.py update).

The Memoization pattern

The **Memoization** pattern is a crucial optimization technique in software development that improves the efficiency of programs by caching the results of expensive function calls. This approach ensures that if a function is called with the same inputs more than once, the cached result is returned, eliminating the need for repetitive and costly computations.

Real-world examples

We can think of calculating Fibonacci numbers as a classic example of the memoization pattern. By storing previously computed values of the sequence, the algorithm avoids recalculating them, which drastically speeds up the computation of higher numbers in the sequence.

Another example is a text search algorithm. In applications dealing with large volumes of text, such as search engines or document analysis tools, caching the results of previous searches means that identical queries can return instant results, significantly improving user experience.

Use cases for the memoization pattern

The memoization pattern can be useful for the following use cases:

1. **Speeding up recursive algorithms**: Memoization transforms recursive algorithms from having a high time complexity. This is particularly beneficial for algorithms such as those calculating Fibonacci numbers.

2. **Reducing computational overhead**: Memoization conserves CPU resources by avoiding unnecessary recalculations. This is crucial in resource-constrained environments or when dealing with high-volume data processing.

3. **Improving application performance**: The direct result of memoization is a noticeable improvement in application performance, making applications feel more responsive and efficient from the user's perspective.

Implementing the memoization pattern

Let's discuss an implementation of the memoization pattern using Python's `functools.lru_cache` decorator. This tool is particularly effective for functions with expensive computations that are called repeatedly with the same arguments. By caching the results, subsequent calls with the same arguments retrieve the result from the cache, significantly reducing execution time.

For our example, we will apply memoization to a classic problem where a recursive algorithm is used: calculating Fibonacci numbers.

We start with the `import` statements we need:

```
from datetime import timedelta
from functools import lru_cache
```

Second, we create a `fibonacci_func1` function that does the Fibonacci numbers computation using recursion (without any caching involved). We will use this for comparison:

```
def fibonacci_func1(n):
    if n < 2:
```

```
        return n
    return fibonacci_func1(n - 1) + fibonacci_func1(n - 2)
```

Third, we define a `fibonacci_func2` function, with the same code, but this one is decorated with `lru_cache`, to enable memoization. What happens here is that the results of the function calls are stored in a cache in memory, and repeated calls with the same arguments fetch results directly from the cache rather than executing the function's code. The code is as follows:

```
@lru_cache(maxsize=None)
def fibonacci_func2(n):
    if n < 2:
        return n
    return fibonacci_func2(n - 1) + fibonacci_func2(n - 2)
```

Finally, we create a `main()` function to test calling both functions using `n=30` as input and measuring the time spent for each execution. The testing code is as follows:

```
def main():
    import time

    n = 30

    start_time = time.time()
    result = fibonacci_func1(n)
    duration = timedelta(time.time() - start_time)
    print(f"Fibonacci_func1({n}) = {result}, calculated in
{duration}")

    start_time = time.time()
    result = fibonacci_func2(n)
    duration = timedelta(time.time() - start_time)
    print(f"Fibonacci_func2({n}) = {result}, calculated in
{duration}")
```

To test the implementation, run the following command: `python ch08/memoization.py`. You should get an output like the following one:

```
Fibonacci_func1(30) = 832040, calculated in 7:38:53.090973
Fibonacci_func2(30) = 832040, calculated in 0:00:02.760315
```

Of course, the durations you get would probably be different than mine, but the duration for the second function, the one using caching, should be less than the one for the function without caching. Also, the difference between both durations should be important.

This was a demonstration that memoization reduces the number of recursive calls needed to calculate Fibonacci numbers, especially for large n values. By reducing the computational overhead, memoization not only speeds up calculations but also conserves system resources, leading to a more efficient and responsive application.

The Lazy Loading pattern

The **Lazy Loading** pattern is a critical design approach in software engineering, particularly useful in optimizing performance and resource management. The idea with lazy loading is to defer the initialization or loading of resources to the moment they are really needed. This way, applications can achieve more efficient resource utilization, reduce initial load times, and enhance the overall user experience.

Real-world examples

Browsing an online art gallery provides a first example. Instead of waiting for hundreds of high-resolution images to load upfront, the website loads only images currently in view. As you scroll, additional images load seamlessly, enhancing your browsing experience without overwhelming your device's memory or network bandwidth.

Another example is an on-demand video streaming service, such as Netflix or YouTube. Such a platform offers an uninterrupted viewing experience by loading videos in chunks. This approach not only minimizes buffering times at the start but also adapts to changing network conditions, ensuring consistent video quality with minimal interruptions.

In applications such as Microsoft Excel or Google Sheets, working with large datasets can be resource-intensive. Lazy loading allows these applications to load only data relevant to your current view or operation, such as a particular sheet or a range of cells. This significantly speeds up operations and reduces memory usage.

Use cases for the lazy loading pattern

We can think of the following performance-related use cases for the lazy loading pattern:

1. **Reducing initial load time**: This is particularly beneficial in web development, where a shorter load time can translate into improved user engagement and retention rates.
2. **Conserving system resources**: In an era of diverse devices, from high-end desktops to entry-level smartphones, optimizing resource usage is crucial for delivering a uniform user experience across all platforms.
3. **Enhancing user experience**: Users expect fast, responsive interactions with software. Lazy loading contributes to this by minimizing waiting times and making applications feel more responsive.

Implementing the lazy loading pattern – lazy attribute loading

Consider an application that performs complex data analysis or generates sophisticated visualizations based on user input. The computation behind this is resource-intensive and time-consuming. Implementing lazy loading, in this case, can drastically improve performance. But for demonstration purposes, we will be less ambitious than the complex data analysis application scenario. We will use a function that simulates an expensive computation and returns a value used for an attribute on a class.

For this lazy loading example, the idea is to have a class that initializes an attribute only when it's accessed for the first time. This approach is commonly used in scenarios where initializing an attribute is resource-intensive, and you want to postpone this process until it's necessary.

We start with the initialization part of the LazyLoadedData class, where we set the _data attribute to None. Here, the expensive data hasn't been loaded yet:

```
class LazyLoadedData:
    def __init__(self):
        self._data = None
```

We add a data() method, decorated with @property, making it act like an attribute (a property) with the added logic for lazy loading. Here, we check if _data is None. If it is, we call the load_data() method:

```
@property
def data(self):
    if self._data is None:
        self._data = self.load_data()
    return self._data
```

We add the load_data() method simulating an expensive operation, using sum(i * i for i in range(100000)). In a real-world scenario, this could involve fetching data from a remote database, performing a complex calculation, or other resource-intensive tasks:

```
def load_data(self):
    print("Loading expensive data...")
    return sum(i * i for i in range(100000))
```

We then add a main() function to test the implementation. We create an instance of the LazyLoadedData class and access the _data attribute twice:

```
def main():
    obj = LazyLoadedData()
    print("Object created, expensive attribute not loaded yet.")

    print("Accessing expensive attribute:")
    print(obj.data)
```

```
    print("Accessing expensive attribute again, no reloading occurs:")
    print(obj.data)
```

To test the implementation, run the `python ch08/lazy_loading/lazy_attribute_loading.py` command. You should get the following output:

```
Object created, expensive attribute not loaded yet.
Accessing expensive attribute:
Loading expensive data...
333328333350000
Accessing expensive attribute again, no reloading occurs:
333328333350000
```

As we can see, on the first access, the expensive data is loaded and stored in `_data`. On subsequent accesses, the data stored is retrieved (from the attribute) without re-performing the expensive operation.

The lazy loading pattern, applied this way, is very useful for improving performance in applications where certain data or computations are needed from time to time but are expensive to produce.

Implementing the lazy loading pattern – using caching

In this second example, we consider a function that calculates the factorial of a number using recursion, which can become quite expensive computationally as the input number grows. While Python's `math` module provides a built-in function for calculating factorials efficiently, implementing it recursively serves as a good example of an expensive computation that could benefit from caching. We will use caching with `lru_cache`, as in the previous section, but this time for the purpose of lazy loading.

We start with importing the modules and functions we need:

```
import time
from datetime import timedelta
from functools import lru_cache
```

Then, we create a `recursive_factorial()` function that calculates the factorial of a number n recursively:

```
def recursive_factorial(n):
    """Calculate factorial (expensive for large n)"""
    if n == 1:
        return 1
    else:
        return n * recursive_factorial(n - 1)
```

Third, we create a `cached_factorial()` function that returns the result of calling `recursive_factorial()` and is decorated with `@lru_cache`. This way, if the function is called again with the same arguments, the result is retrieved from the cache instead of being recalculated, significantly reducing computation time:

```python
@lru_cache(maxsize=128)
def cached_factorial(n):
    return recursive_factorial(n)
```

We create a `main()` function as usual for testing the functions. We call the non-cached function, and then we call the `cached_factorial` function twice, showing the computation time for each case. The code is as follows:

```python
def main():
    # Testing the performance
    n = 20

    # Without caching
    start_time = time.time()
    print(f"Recursive factorial of {n}: {recursive_factorial(n)}")
    duration = timedelta(time.time() - start_time)
    print(f"Calculation time without caching: {duration}.")

    # With caching
    start_time = time.time()
    print(f"Cached factorial of {n}: {cached_factorial(n)}")
    duration = timedelta(time.time() - start_time)
    print(f"Calculation time with caching: {duration}.")

    start_time = time.time()
    print(f"Cached factorial of {n}, repeated: {cached_factorial(n)}")
    duration = timedelta(time.time() - start_time)
    print(f"Second calculation time with caching: {duration}.")
```

To test the implementation, run the `python ch08/lazy_loading/lazy_loading_with_caching.py` command. You should get the following output:

```
Recursive factorial of 20: 2432902008176640000
Calculation time without caching: 0:00:04.840851
Cached factorial of 20: 2432902008176640000
Calculation time with caching: 0:00:00.865173
Cached factorial of 20, repeated: 2432902008176640000
Second calculation time with caching: 0:00:00.350189
```

You will notice the time taken for the initial calculation of the factorial without caching, then the time with caching, and finally, the time for a repeated calculation with caching.

Also, `lru_cache` is inherently a memoization tool, but it can be adapted and used in cases where, for example, there are expensive initialization processes that need to be executed only when required and not make the application slow. In our example, we used factorial computation to simulate such expensive processes.

If you are asking yourself what is the difference from memoization, the answer is that the context in which caching is used here is for managing resource initialization.

Summary

Throughout this chapter, we have explored patterns that developers can use to enhance the efficiency and scalability of applications.

The cache-aside pattern teaches us how to manage cache effectively, ensuring data is fetched and stored in a manner that optimizes performance and consistency, particularly in environments with dynamic data sources.

The memoization pattern demonstrates the power of caching function results to speed up applications by avoiding redundant computations. This pattern is beneficial for expensive, repeatable operations and can dramatically improve the performance of recursive algorithms and complex calculations.

Finally, the lazy loading pattern emphasizes delaying the initialization of resources until they are needed. This approach not only improves the startup time of applications but also reduces memory overhead, making it ideal for resource-intensive operations that may not always be necessary for the user's interactions.

In the next chapter, we are going to discuss patterns that govern distributed systems.

9

Distributed Systems Patterns

As technology evolves and the demand for scalable and resilient systems increases, understanding the fundamental patterns that govern distributed systems becomes paramount.

From managing communication between nodes to ensuring **fault tolerance** (**FT**) and consistency, this chapter explores essential design patterns that empower developers to architect robust distributed systems. Whether you're building microservices or implementing cloud-native applications, mastering these patterns will equip you with the tools to tackle the complexities of distributed computing effectively.

In this chapter, we're going to cover the following main topics:

- The Throttling pattern
- The Retry pattern
- The Circuit Breaker pattern
- Other distributed systems patterns

Technical requirements

See the requirements presented in *Chapter 1*. The additional technical requirements for the code discussed in this chapter are the following:

- Install Flask and Flask-Limiter, using `python -m pip install flask flask-limiter`
- Install PyBreaker, using `python -m pip install pybreaker`

The Throttling pattern

Throttling is an important pattern we may need to use in today's applications and APIs. In this context, throttling means controlling the rate of requests a user (or a client service) can send to a given service or API in a given amount of time, to protect the resources of the service from being overused. For example, we may limit the number of user requests for an API to 1,000 per day. Once that limit is reached, the next request is handled by sending an error message with the 429 HTTP status code to the user with a message saying that there are too many requests.

There are many things to understand about throttling, including which limiting strategy and algorithm one may use and measuring how the service is used. You can find technical details about the Throttling pattern in the catalog of cloud design patterns by Microsoft (`https://learn.microsoft.com/en-us/azure/architecture/patterns/throttling`).

Real-world examples

There are a lot of examples of throttling in real life, such as the following:

- **Highway traffic management**: Traffic lights or speed limits regulate the flow of vehicles on a highway

- **Water faucet**: Adjusting the flow of water from a faucet

- **Concert ticket sales**: When tickets for a popular concert go on sale, the website may limit the number of tickets each user can purchase at once to prevent the server from crashing due to a sudden surge in demand

- **Electricity usage**: Some utility companies offer plans where customers pay different rates based on their electricity usage during peak and off-peak hours

- **Buffet line**: In a buffet, customers may be limited to taking only one plate of food at a time to ensure that everyone has a fair chance to eat and to prevent food wastage

We also have examples of pieces of software that help implement throttling:

- `django-throttle-requests` (`https://github.com/sobotklp/django-throttle-requests`) is a framework for implementing application-specific rate-limiting middleware for Django projects

- Flask-Limiter (`https://flask-limiter.readthedocs.io/en/stable/`) provides rate-limiting features to Flask routes

Use cases for the Throttling pattern

This pattern is recommended when you need to ensure your system continuously delivers the service as expected, when you need to optimize the cost of usage of the service, or when you need to handle bursts in activity.

In practice, you may implement the following rules:

- Limit the number of total requests to an API as N/day (for example, N=1000)
- Limit the number of requests to an API as N/day from a given IP address, or from a given country or region
- Limit the number of reads or writes for authenticated users

In addition to the rate-limiting cases, it can be used for *resource allocation*, ensuring fair distribution of resources among multiple clients.

Implementing the Throttling pattern

Before diving into an implementation example, you need to know that there are several types of throttling, among which are Rate-Limit, IP-level Limit (based on a list of whitelisted IP addresses, for example), and Concurrent Connections Limit, to only cite those three. The first two are relatively easy to experiment with. We will focus on the first one here.

Let's see an example of rate-limit-type throttling using a minimal web application developed using Flask and its Flask-Limiter extension.

We start with the imports we need for the example:

```
from flask import Flask
from flask_limiter import Limiter
from flask_limiter.util import get_remote_address
```

As is usual with Flask, we set up the Flask application with the following two lines:

```
app = Flask(__name__)
```

We then define the Limiter instance; we create it by passing a key function, `get_remote_address` (which we imported), the application object, the default limits values, and other parameters, as follows:

```
limiter = Limiter(
    get_remote_address,
    app=app,
    default_limits=["100 per day", "10 per hour"],
    storage_uri="memory://",
```

```
        strategy="fixed-window",
    )
```

Based on that, we can define a route for the /limited path, which will be rate-limited using the default limits, as follows:

```
@app.route("/limited")
def limited_api():
    return "Welcome to our API!"
```

We also add the definition for a route for the /more_limited path. In this case, we decorate the function with @limiter.limit("2/minute") to ensure a rate limit of two requests per minute. The code is as follows:

```
@app.route("/more_limited")
@limiter.limit("2/minute")
def more_limited_api():
    return "Welcome to our expensive, thus very limited, API!"
```

Finally, we add the snippet that is conventional for Flask applications:

```
if __name__ == "__main__":
    app.run(debug=True)
```

To test this example, run the file (ch09/throttling_flaskapp.py) using the python ch09/throttling_flaskapp.py command. You would get the usual output for a Flask application that is starting:

```
* Serving Flask app 'throttling_flaskapp'
* Debug mode: on
WARNING: This is a development server. Do not use it in a production deployment. Use a production WSGI server instead.
* Running on http://127.0.0.1:5000
Press CTRL+C to quit
* Restarting with stat
* Debugger is active!
* Debugger PIN: 619-166-428
```

Figure 9.1 – throttling_flaskapp: Flask application example startup

Then, if you point your browser to http://127.0.0.1:5000/limited, you will see the welcome content displayed on the page, as follows:

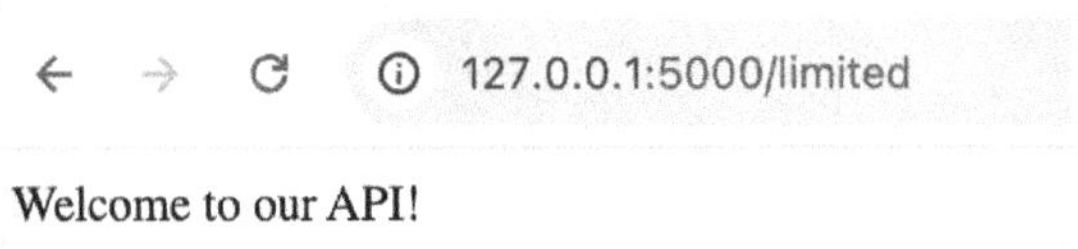

← → C ⓘ 127.0.0.1:5000/limited

Welcome to our API!

Figure 9.2 – Response to the /limited endpoint in the browser

It gets interesting if you keep hitting the **Refresh** button. The 10th time, the page content will change and show you a **Too Many Requests** error message, as shown in the following screenshot:

Figure 9.3 – Too many requests on the /limited endpoint

Let's not stop here. Remember – there is a second route in the code, `/more_limited`, with a specific limit of two requests per minute. To test that second route, point your browser to `http://127.0.0.1:5000/more_limited`. You will see new welcome content displayed on the page, as follows:

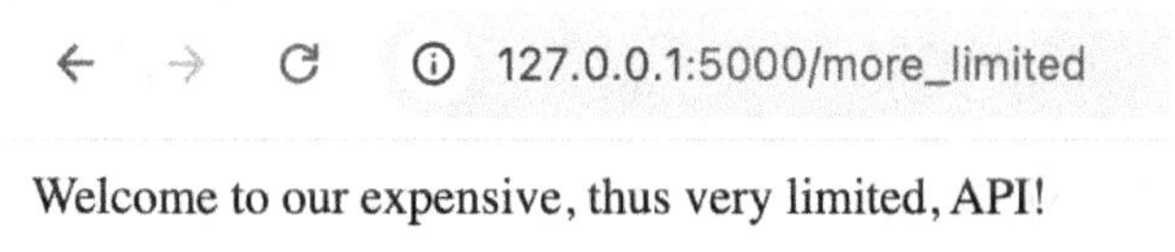

Figure 9.4 – Response to the /more_limited endpoint in the browser

If we hit the **Refresh** button and do it more than twice in a window of 1 minute, we get another **Two Many Requests** message, as shown in the following screenshot:

Figure 9.5 – Too many requests on the /more_limited endpoint

Also, looking at the console where the Flask server is running, you will notice the mention of each HTTP request received and the status code of the response the application sent. It should look like the following screenshot:

```
127.0.0.1 - - [23/Apr/2024 18:23:44] "GET /limited HTTP/1.1" 200 -
127.0.0.1 - - [23/Apr/2024 18:26:42] "GET /limited HTTP/1.1" 200 -
127.0.0.1 - - [23/Apr/2024 18:26:47] "GET /limited HTTP/1.1" 200 -
127.0.0.1 - - [23/Apr/2024 18:26:50] "GET /limited HTTP/1.1" 200 -
127.0.0.1 - - [23/Apr/2024 18:26:53] "GET /limited HTTP/1.1" 200 -
127.0.0.1 - - [23/Apr/2024 18:26:54] "GET /limited HTTP/1.1" 200 -
127.0.0.1 - - [23/Apr/2024 18:26:56] "GET /limited HTTP/1.1" 200 -
127.0.0.1 - - [23/Apr/2024 18:26:57] "GET /limited HTTP/1.1" 200 -
127.0.0.1 - - [23/Apr/2024 18:26:57] "GET /limited HTTP/1.1" 200 -
127.0.0.1 - - [23/Apr/2024 18:26:58] "GET /limited HTTP/1.1" 200 -
127.0.0.1 - - [23/Apr/2024 18:26:59] "GET /limited HTTP/1.1" 429 -
127.0.0.1 - - [23/Apr/2024 20:16:39] "GET /more_limited HTTP/1.1" 200 -
127.0.0.1 - - [23/Apr/2024 20:16:57] "GET /more_limited HTTP/1.1" 200 -
127.0.0.1 - - [23/Apr/2024 20:16:59] "GET /more_limited HTTP/1.1" 429 -
```

Figure 9.6 – Flask server console: Responses to the HTTP requests

There are many possibilities for rate-limit-type throttling in a Flask application using the Flask-Limiter extension, as you can see on the documentation page of the module. The reader can find more information on the documentation page on how to use different strategies and storage backends such as Redis for a specific implementation.

The Retry pattern

Retrying is an approach that is increasingly needed in the context of distributed systems. Think about microservices or cloud-based infrastructures where components collaborate with each other but are not developed or deployed/operated by the same teams and parties.

In its daily operation, parts of a cloud-native application may experience what are called transient faults or failures, meaning some mini-issues that can look like bugs but are not due to your application itself; rather, they are due to some constraints outside of your control such as the networking or the external server/service performance. As a result, your application may malfunction (at least, that could be the perception of your users) or even hang in some places. The answer to the risk of such failures is to put in place some retry logic so that we pass through the issue by calling the service again, maybe immediately or after some wait time (such as a few seconds).

Real-world examples

There are examples of the Retry pattern (or analogies) in our daily life, such as the following:

- **Making a phone call**: Imagine you're trying to reach a friend on the phone, but the call doesn't go through because their line is busy or there's a network issue. Instead of giving up immediately, you retry dialing their number after a short delay.

- **Withdrawing money from an ATM**: Imagine you go to an ATM to withdraw cash, but due to a temporary issue such as network congestion or connectivity problems, the transaction fails, and the machine displays an error message. Instead of giving up on getting cash, you wait a moment and try the transaction again. This time, the transaction may go through successfully, allowing you to withdraw the money you need.

There are also many tools or techniques that we can consider as examples in the software realm since they help implement the Retry pattern, such as the following:

- In Python, the Retrying library (`https://github.com/rholder/retrying`) is available to simplify the task of adding retry behavior to our functions

- The Pester library (`https://github.com/sethgrid/pester`) for Go developers

Use cases for the Retry pattern

This pattern is recommended to alleviate the impact of identified transient failures while communicating with an external component or service, due to network failure or server overload.

Note that the retrying approach is not recommended for handling failures such as internal exceptions caused by errors in the application logic itself. Also, we must analyze the response from the external service. If the application experiences frequent busy faults, it's often a sign that the service being accessed has a scaling issue that should be addressed.

We can relate retrying to the microservices architecture, where services often communicate over the network. The Retry pattern ensures that transient failures don't cause the entire system to fail.

Another type of use case is *data synchronization*. When syncing data between two systems, retries can handle the temporary unavailability of one system.

Implementing the Retry pattern

In this example, we'll implement the Retry pattern for a database connection. We'll use a decorator to handle the retry mechanism.

We start with the `import` statements for the example, as follows:

```
import logging
import random
import time
```

We then add configuration for logging, which will help for observability when using the code:

```
logging.basicConfig(level=logging.DEBUG)
```

We add our function that will support the decorator to automatically retry the execution of the decorated function up to the number of attempts specified, as follows:

```
def retry(attempts):
    def decorator(func):
```

```
        def wrapper(*args, **kwargs):
            for _ in range(attempts):
                try:
                    logging.info("Retry happening")
                    return func(*args, **kwargs)
                except Exception as e:
                    time.sleep(1)
                    logging.debug(e)
            return "Failure after all attempts"
        return wrapper

    return decorator
```

Then, we add the `connect_to_database()` function, which simulates a database connection. It is decorated by the `@retry` decorator. We want the decorator to automatically retry the connection up to three times if it fails:

```
@retry(attempts=3)
def connect_to_database():
    if random.randint(0, 1):
        raise Exception("Temporary Database Error")
    return "Connected to Database"
```

Finally, to make it convenient to test our implementation, we add the following testing code:

```
if __name__ == "__main__":
    for i in range(1, 6):
        logging.info(f"Connection attempt #{i}")
        print(f"--> {connect_to_database()}")
```

To test the example, run the following command:

```
python ch09/retry/retry_database_connection.py
```

You should get an output like the following:

```
INFO:root:Connection attempt #1
INFO:root:Retry happening
--> Connected to Database
INFO:root:Connection attempt #2
INFO:root:Retry happening
DEBUG:root:Temporary Database Error
INFO:root:Retry happening
```

```
DEBUG:root:Temporary Database Error
INFO:root:Retry happening
DEBUG:root:Temporary Database Error
--> Failure after all attempts
INFO:root:Connection attempt #3
INFO:root:Retry happening
--> Connected to Database
INFO:root:Connection attempt #4
INFO:root:Retry happening
--> Connected to Database
INFO:root:Connection attempt #5
INFO:root:Retry happening
DEBUG:root:Temporary Database Error
INFO:root:Retry happening
DEBUG:root:Temporary Database Error
INFO:root:Retry happening
DEBUG:root:Temporary Database Error
--> Failure after all attempts
```

So, when a temporary database error occurs, a retry happens. Several retry attempts may occur, until three. Once three unsuccessful retry attempts have occurred, the outcome is the failure of the operation.

Overall, the Retry pattern is a viable way to handle this type of use case involved with distributed systems, and a few errors (four database errors in our example) may mean that there is a more permanent or problematic bug that should be fixed.

The Circuit Breaker pattern

One approach to FT involves retries, as we have just seen. But, when a failure due to communication with an external component is likely to be long-lasting, using a retry mechanism can affect the responsiveness of the application. We might be wasting time and resources trying to repeat a request that's likely to fail. This is where another pattern can be useful: the Circuit Breaker pattern.

With the Circuit Breaker pattern, you wrap a fragile function call, or an integration point with an external service, in a special (circuit breaker) object, which monitors for failures. Once the failures reach a certain threshold, the circuit breaker trips and all subsequent calls to the circuit breaker return with an error, without the protected call being made at all.

Real-world examples

In life, we can think of a water or electricity distribution circuit where a circuit breaker plays an important role.

In software, a circuit breaker is used in the following examples:

- **E-commerce checkout**: If the payment gateway is down, the circuit breaker can halt further payment attempts, preventing system overload
- **Rate-limited APIs**: When an API has reached its rate limit, a circuit breaker can stop additional requests to avoid penalties

Use cases for the Circuit Breaker pattern

As already said, the Circuit Breaker pattern is recommended when you need a component from your system to be fault-tolerant to long-lasting failures when communicating with an external component, service, or resource. Next, we will understand how it addresses such use cases.

Implementing the Circuit Breaker pattern

Let's say you want to use a circuit breaker on a flaky function, a function that is fragile, for example, due to the networking environment it depends on. We are going to use the `pybreaker` library (`https://pypi.org/project/pybreaker/`) to show an example of implementing the Circuit Breaker pattern.

Our implementation is an adaptation of a nice script I found in this repository: `https://github.com/veltra/pybreaker-playground`. Let's go through the code.

We start with our imports, as follows:

```
import pybreaker
from datetime import datetime
import random
from time import sleep
```

Let's define our circuit breaker to automatically open the circuit after five consecutive failures in that function. We need to create an instance of the `pybreaker.CircuitBreaker` class, as follows:

```
breaker = pybreaker.CircuitBreaker(fail_max=2, reset_timeout=5)
```

Then, we create our version of the function to simulate fragile calls. We use the decorator syntax to protect things, so the new function is as follows:

```
@breaker
def fragile_function():
    if not random.choice([True, False]):
        print(" / OK", end="")
    else:
        print(" / FAIL", end="")
        raise Exception("This is a sample Exception")
```

Finally, here's the main part of the code, with the `main()` function:

```
def main():
    while True:
        print(datetime.now().strftime("%Y-%m-%d %H:%M:%S"), end="")

        try:
            fragile_function()
        except Exception as e:
            print(" / {} {}".format(type(e), e), end="")
        finally:
            print("")
        sleep(1)
```

Calling this script by running the `python ch09/circuit_breaker.py` command produces the following output:

```
2024-04-23 22:48:54 / OK
2024-04-23 22:48:55 / FAIL / <class 'Exception'> This is a sample Exception
2024-04-23 22:48:56 / OK
2024-04-23 22:48:57 / FAIL / <class 'Exception'> This is a sample Exception
2024-04-23 22:48:58 / FAIL / <class 'pybreaker.CircuitBreakerError'> Failures threshold reached, circuit breaker opened
2024-04-23 22:48:59 / <class 'pybreaker.CircuitBreakerError'> Timeout not elapsed yet, circuit breaker still open
2024-04-23 22:49:00 / <class 'pybreaker.CircuitBreakerError'> Timeout not elapsed yet, circuit breaker still open
2024-04-23 22:49:01 / <class 'pybreaker.CircuitBreakerError'> Timeout not elapsed yet, circuit breaker still open
2024-04-23 22:49:02 / <class 'pybreaker.CircuitBreakerError'> Timeout not elapsed yet, circuit breaker still open
2024-04-23 22:49:03 / FAIL / <class 'pybreaker.CircuitBreakerError'> Trial call failed, circuit breaker opened
2024-04-23 22:49:04 / <class 'pybreaker.CircuitBreakerError'> Timeout not elapsed yet, circuit breaker still open
2024-04-23 22:49:05 / <class 'pybreaker.CircuitBreakerError'> Timeout not elapsed yet, circuit breaker still open
2024-04-23 22:49:06 / <class 'pybreaker.CircuitBreakerError'> Timeout not elapsed yet, circuit breaker still open
2024-04-23 22:49:07 / <class 'pybreaker.CircuitBreakerError'> Timeout not elapsed yet, circuit breaker still open
2024-04-23 22:49:08 / OK
2024-04-23 22:49:09 / FAIL / <class 'Exception'> This is a sample Exception
2024-04-23 22:49:10 / OK
2024-04-23 22:49:11 / OK
2024-04-23 22:49:12 / FAIL / <class 'Exception'> This is a sample Exception
2024-04-23 22:49:13 / FAIL / <class 'pybreaker.CircuitBreakerError'> Failures threshold reached, circuit breaker opened
2024-04-23 22:49:14 / <class 'pybreaker.CircuitBreakerError'> Timeout not elapsed yet, circuit breaker still open
2024-04-23 22:49:15 / <class 'pybreaker.CircuitBreakerError'> Timeout not elapsed yet, circuit breaker still open
2024-04-23 22:49:16 / <class 'pybreaker.CircuitBreakerError'> Timeout not elapsed yet, circuit breaker still open
```

Figure 9.7 – Output of our program using a circuit breaker

By closely looking at the output, we can see that the circuit breaker does its job as expected: when it is open, all `fragile_function()` calls fail immediately (since they raise the `CircuitBreakerError` exception) without any attempt to execute the intended operation. And, after a timeout of 5 seconds, the circuit breaker will allow the next call to go through. If that call succeeds, the circuit is closed; if it fails, the circuit is opened again until another timeout elapses.

Other distributed systems patterns

There are many more distributed systems patterns than the ones we covered here. Among the other patterns developers and architects can use are the following:

- **Command and Query Responsibility Segregation (CQRS)**: This pattern separates the responsibilities for reading and writing data, allowing for optimized data access and scalability by tailoring data models and operations to specific use cases

- **Two-Phase Commit**: This distributed transaction protocol ensures atomicity and consistency across multiple participating resources by coordinating a two-phase commit process, involving a *prepare* phase followed by a *commit* phase

- **Saga**: A saga is a sequence of local transactions that together form a distributed transaction, providing a compensating mechanism to maintain consistency in the face of partial failures or aborted transactions

- **Sidecar**: The Sidecar pattern involves deploying additional helper services alongside primary services to enhance functionality, such as adding monitoring, logging, or security features without directly modifying the main application

- **Service Registry**: This pattern centralizes the management and discovery of services within a distributed system, allowing services to dynamically register and discover each other, facilitating communication and scalability

- **Bulkhead**: Inspired by ship design, the Bulkhead pattern partitions resources or components within a system to isolate failures and prevent cascading failures from impacting other parts of the system, thereby enhancing FT and resilience

Each of these patterns addresses specific challenges inherent in distributed systems, offering strategies and best practices for architects and developers to design robust and scalable solutions capable of operating in dynamic and unpredictable environments.

Summary

In this chapter, we delved into the intricacies of distributed systems patterns, focusing on the Throttling, Retry, and Circuit Breaker patterns. These patterns are essential for building robust, fault-tolerant, and efficient distributed systems.

The skills you've acquired in this chapter will significantly contribute to your ability to design and implement distributed systems that can handle transient failures, service interruptions, and high loads.

The section about the Throttling pattern equipped you with the tools to manage service load and resource allocation effectively.

By understanding how to implement the Retry pattern, you've gained the skills to make your operations more reliable.

And, finally, the Circuit Breaker pattern taught you how to build fault-tolerant systems that can gracefully handle failures.

As we wrap up this chapter, it's crucial to remember that these patterns are not isolated solutions but pieces of a larger puzzle. They often work best when combined and tailored to fit the specific needs and constraints of your system. The key takeaway is to understand the underlying principles so that you can adapt them to create a resilient and efficient distributed system.

Lastly, we briefly presented some other distributed systems patterns, which we cannot cover in this book.

In the next chapter, we will focus on patterns for testing.

10
Patterns for Testing

In the previous chapters, we covered architectural patterns and patterns for specific use cases such as concurrency or performance.

In this chapter, we will explore design patterns that are particularly useful for testing. These patterns help in isolating components, making tests more reliable, and promoting code reusability.

In this chapter, we're going to cover the following main topics:

- The Mock Object pattern
- The Dependency Injection pattern

Technical requirements

See the requirements presented in *Chapter 1*.

The Mock Object pattern

The **Mock Object** pattern is a powerful tool for isolating components during testing by simulating their behavior. Mock objects help create controlled testing environments and verify interactions between components.

The Mock Object pattern provides three features:

1. **Isolation**: Mocks isolate the unit of code being tested, ensuring that tests run in a controlled environment where dependencies are predictable and do not have external side effects.

2. **Behavior verification**: By using mock objects, you can verify that certain behaviors happen during a test, such as method calls or property accesses.

3. **Simplification**: They simplify the setup of tests by replacing complex real objects that might require significant setup of their own.

> **Comparison with stubs**
>
> Stubs also replace real implementations but are used only to provide indirect input to the code under test. Mocks, by contrast, can also verify interactions, making them more flexible in many testing scenarios.

Real-world examples

We can think of the following analog concepts or tools in the real world:

- A flight simulator, which is a tool designed to replicate the experience of flying an actual airplane. It allows pilots to learn how to handle various flight scenarios in a controlled and safe environment.

- A **cardiopulmonary resuscitation** (CPR) dummy, which is used to teach students how to perform CPR effectively. It simulates the human body to provide a realistic yet controlled setting for learning.

- A crash test dummy, which is used by car manufacturers to simulate human reactions to vehicle collisions. It provides valuable data on the impacts and safety features of a car without putting actual human lives at risk.

Use cases for the Mock Object pattern

In **unit testing**, mock objects are used to replace complex, unreliable, or unavailable dependencies of the code being tested. This allows developers to focus solely on the unit itself rather than its interactions with external systems. For example, when testing a service that fetches data from an API, a mock object can simulate the API by returning predefined responses, ensuring that the service can handle various data scenarios or errors without needing to interact with the actual API.

While similar to unit testing, **integration testing** with mock objects focuses on the interaction between components rather than individual units. Mocks can be used to simulate components that have not been developed yet or are too costly to involve in every test. For example, in a microservices architecture, a mock can represent a service that is under development or temporarily unavailable, allowing other services to be tested in terms of how they integrate and communicate with it.

The Mock Object pattern is also useful for **behavior verification**. This use case involves verifying that certain interactions between objects occur as expected. Mock objects can be programmed to expect specific calls, arguments, and even order of interactions, which makes them powerful tools for behavioral testing; for example, testing whether a controller, in a **Model View Controller** (MVC) architecture, correctly calls authentication and logging services before processing a user request. Mocks can verify that the controller makes the right calls in the right order, such as checking credentials before attempting to log the request.

Implementing the Mock Object pattern

Imagine we have a function that logs messages to a file. We can mock the file-writing mechanism to ensure our logging function writes the expected content to the log without writing to a file. Let's see how this can be implemented using Python's `unittest` module.

First, we import what we need for the example:

```
import unittest
from unittest.mock import mock_open, patch
```

Then, we create a class representing a simple logger that writes messages to a file specified during initialization:

```
class Logger:
    def __init__(self, filepath):
        self.filepath = filepath

    def log(self, message):
        with open(self.filepath, "a") as file:
            file.write(f"{message}\n")
```

Next, we create a test case class that inherits from the `unittest.TestCase` class, as usual. In this class, we need the `test_log()` method to test the logger's `log()` method, as follows:

```
class TestLogger(unittest.TestCase):
    def test_log(self):
        msg = "Hello, logging world!"
```

Next, we are going to mock the Python built-in `open()` function directly within the test scope. Mocking the function is done using `unittest.mock.patch()`, which temporarily replaces the target object, `builtins.open`, with a mock object (the result of calling `mock_open()`). With the context manager we get from calling the `unittest.mock.patch()` function, we create a `Logger` object and call its `.log()` method, which should trigger the `open()` function:

```
        m_open = mock_open()

        with patch("builtins.open", m_open):
            logger = Logger("dummy.log")
            logger.log(msg)
```

About builtins

According to Python documentation, the `builtins` module provides direct access to all built-in identifiers of Python; for example, `builtins.open` is the full name for the `open()` built-in function. See `https://docs.python.org/3/library/builtins.html`.

About mock_open

When you call `mock_open()`, it returns a Mock object that is configured to behave like the built-in `open()` function. This mock is set up to simulate file operations such as reading and writing.

About unittest.mock.patch

It is used to replace objects with mocks during testing. Its arguments include `target` to specify the object to replace, and optional arguments: `new` for an optional replacement object, `spec` and `autospec` to limit the mock to the real object's attributes for accuracy, `spec_set` for a stricter attribute specification, `side_effect` to define conditional behavior or exceptions, `return_value` for setting a fixed response, and `wraps` to allow the original object's behavior while modifying certain aspects. These options enable precise control and flexibility in testing scenarios.

Now, we check that the log file was opened correctly, which we do using two verifications. For the first one, we use the `assert_called_once_with()` method on the mock object, to check that the `open()` function was called with the expected parameters. For the second one, we need more tricks from `unittest.mock.mock_open`; our m_open mock object, which was obtained by calling the `mock_open()` function, is also a callable object that behaves like a factory for creating new mock file handles each time it's called. We use that to obtain a new file handle, and then we use `assert_called_once_with()` on the `write()` method call on that file handle, which helps us check if the `write()` method was called with the correct message. This part of the test function is as follows:

```
m_open.assert_called_once_with(
    "dummy.log", "a"
)
m_open().write.assert_called_once_with(
    f"{msg}\n"
)
```

Finally, we call `unitest.main()`:

```
if __name__ == "__main__":
    unittest.main()
```

To execute the example (in the `ch10/mock_object.py` file), as usual, run the following command:

```
python ch10/mock_object.py
```

You should get an output like the following:

```
.
--------------------------------------------------------------
```

```
Ran 1 test in 0.012s
```

```
OK
```

That was a quick demonstration of using mocking to simulate parts of a system within a unit test. We can see that this approach isolates side effects (that is, file I/O), ensuring that the unit tests do not create or require actual files. It allows testing the internal behavior of the class without altering the class structure for testing purposes.

The Dependency Injection pattern

The Dependency Injection pattern involves passing the dependencies of a class as external entities rather than creating them within the class. This promotes loose coupling, modularity, and testability.

Real-world examples

We come across the following examples in real life:

- **Electrical appliances and power outlets**: Various electrical appliances can be plugged into different power outlets to use electricity without needing direct and permanent wiring
- **Lenses on cameras**: A photographer can change lenses on a camera to suit different environments and needs without changing the camera itself
- **Modular train systems**: In a modular train system, individual cars (such as sleeper, diner, or baggage cars) can be added or removed depending on the needs of each journey

Use cases for the Dependency Injection pattern

In web applications, injecting database connection objects into components such as repositories or services enhances modularity and maintainability. This practice allows for an easy switch between different database engines or configurations without the need to alter the component's code directly. It also significantly simplifies the process of unit testing by enabling the injection of mock database connections, thereby testing various data scenarios without affecting the live databases.

Another type of use case is managing configuration settings across various environments (development, testing, production, and so on). By dynamically injecting settings into modules, **dependency injection (DI)** reduces coupling between the modules and their configuration sources. This flexibility makes it easier to manage and switch environments without needing extensive reconfiguration. In unit testing, this means you can inject specific settings to test how modules perform under different configurations, ensuring robustness and functionality.

Implementing the Dependency Injection pattern – using a mock object

In this first example, we'll create a simple scenario where a `WeatherService` class depends on a `WeatherApiClient` interface to fetch weather data. For the example's unit test code, we will inject a mock version of this API client.

We start by defining the interface any weather API client implementation should conform to, using Python's `Protocol` feature:

```python
from typing import Protocol


class WeatherApiClient(Protocol):
    def fetch_weather(self, location):
        """Fetch weather data for a given location"""
        ...
```

Then, we add a `RealWeatherApiClient` class that implements that interface and that would interact with our weather service. In a real scenario, in the provided `fetch_weather()` method, we would perform a call to a weather service, but we want to keep the example simple and focus on the main concepts of this chapter; so, we provide a simulation, simply returning a string that represents the weather data result. The code is as follows:

```python
class RealWeatherApiClient:
    def fetch_weather(self, location):
        return f"Real weather data for {location}"
```

Next, we create a weather service, which uses an object that implements the `WeatherApiClient` interface to fetch weather data:

```python
class WeatherService:
    def __init__(self, weather_api: WeatherApiClient):
        self.weather_api = weather_api

    def get_weather(self, location):
        return self.weather_api.fetch_weather(location)
```

Finally, we are ready to inject the API client's dependency through the `WeatherService` constructor. We add code that helps manually test the example, using the real service, as follows:

```python
if __name__ == "__main__":
    ws = WeatherService(RealWeatherApiClient())
    print(ws.get_weather("Paris"))
```

This part of our example (in the `ch10/dependency_injection/di_with_mock.py` file) can be manually tested by using the following command:

```
python ch10/dependency_injection/di_with_mock.py
```

You should get the following output:

```
Real weather data for Paris
```

Since the interesting part of our example is about unit testing, let's add that part (in a second file, ch10/dependency_injection/test_di_with_mock.py).

First, we import the `unittest` module, as well as the `WeatherService` class (from our di_with_mock module), as follows:

```
import unittest
from di_with_mock import WeatherService
```

Then, we create a mock version of the weather API client implementation that will be useful for unit testing, simulating responses without making real API calls:

```
class MockWeatherApiClient:
    def fetch_weather(self, location):
        return f"Mock weather data for {location}"
```

Next, we write the test case class, with a test function. In that function, we inject the mock API client instead of the real API client, passing it to the `WeatherService` constructor, as follows:

```
class TestWeatherService(unittest.TestCase):
    def test_get_weather(self):
        mock_api = MockWeatherApiClient()
        weather_service = WeatherService(mock_api)
        self.assertEqual(
            weather_service.get_weather("Anywhere"),
            "Mock weather data for Anywhere",
        )
```

We finish by adding the usual lines for executing unit tests when the file is interpreted by Python:

```
if __name__ == "__main__":
    unittest.main()
```

Executing this part of the example (in the ch10/dependency_injection/test_di_with_mock.py file), using the python ch10/dependency_injection/test_di_with_mock.py command, gives the following output:

```
.
----------------------------------------------------------------
```

```
Ran 1 test in 0.000s
```

```
OK
```

The test with the dependency injected using a mock object succeeded.

Through this example, we were able to see that the `WeatherService` class doesn't need to know whether it's using a real or a mock API client, making the system more modular and easier to test.

Implementing the Dependency Injection pattern – using a decorator

It is also possible to use decorators for DI, which simplifies the injection process. Let's see a simple example demonstrating how to do that, where we'll create a notification system that can send notifications through different channels (for example, email or SMS). The first part of the example will show the result based on manual testing, and the second part will provide unit tests.

First, we define a `NotificationSender` interface, outlining the methods any notification sender should have:

```
from typing import Protocol

class NotificationSender(Protocol):
    def send(self, message: str):
        """Send a notification with the given message"""
        ...
```

Then, we implement two specific notification senders: the `EmailSender` class implements sending a notification using email, and the `SMSSender` class implements sending using SMS. This part of the code is as follows:

```
class EmailSender:
    def send(self, message: str):
        print(f"Sending Email: {message}")


class SMSSender:
    def send(self, message: str):
        print(f"Sending SMS: {message}")
```

We also define a notification service class, `NotificationService`, with a class attribute sender and a `.notify()` method, which takes in a message and calls `.send()` on the provided sender object to send the message, as follows:

```
class NotificationService:
    sender: NotificationSender = None
```

```
    def notify(self, message):
        self.sender.send(message)
```

What is missing is the decorator that will operate the DI, to provide the specific sender object to be used. We create our decorator to decorate the `NotificationService` class for injecting the sender. It will be used by calling `@inject_sender(EmailSender)` if we want to inject the email sender, or `@inject_sender(SMSSender)` if we want to inject the SMS sender. The code for the decorator is as follows:

```
def inject_sender(sender_cls):
    def decorator(cls):
        cls.sender = sender_cls()
        return cls

    return decorator
```

Now, if we come back to the notification service's class, the code would be as follows:

```
@inject_sender(EmailSender)
class NotificationService:
    sender: NotificationSender = None

    def notify(self, message):
        self.sender.send(message)
```

Finally, we can instantiate the `NotificationService` class in our application and notify a message for testing the implementation, as follows:

```
if __name__ == "__main__":
    service = NotificationService()
    service.notify("Hello, this is a test notification!")
```

That first part of our example (in the `ch10/dependency_injection/di_with_decorator.py` file) can be manually tested by using the following command:

```
python ch10/dependency_injection/di_with_decorator.py
```

You should get the following output:

```
Sending Email: Hello, this is a test notification!
```

If you change the decorating line, replace the `EmailSender` class with `SMSSender`, and rerun that command, you will get the following output:

```
Sending SMS: Hello, this is a test notification!
```

That shows the DI is effective.

Next, we want to write unit tests for that implementation. We could use the mocking technique, but to see other ways, we are going to use the stub classes approach. The stubs manually implement the dependency interfaces and include additional mechanisms to verify that methods have been called correctly. Let's start by importing what we need:

```
import unittest

from di_with_decorator import (
    NotificationSender,
    NotificationService,
    inject_sender,
)
```

Then, we create stub classes that implement the `NotificationSender` interface. These classes will help record calls to their `send()` method, using the `messages_sent` attribute on their instances, allowing us to check whether the correct methods were called during the test. Both stub classes are as follows:

```
class EmailSenderStub:
    def __init__(self):
        self.messages_sent = []

    def send(self, message: str):
        self.messages_sent.append(message)


class SMSSenderStub:
    def __init__(self):
        self.messages_sent = []

    def send(self, message: str):
        self.messages_sent.append(message)
```

Next, we are going to use both stubs in our test case to verify the functionality of `NotificationService`. In the test function, `test_notify_with_email`, we create an instance of `EmailSenderStub`, inject that stub into the service, send a notification message, and then verify that the message was sent by the email stub. That part of the code is as follows:

```
class TestNotifService(unittest.TestCase):
    def test_notify_with_email(self):
        email_stub = EmailSenderStub()

        service = NotificationService()
```

```
        service.sender = email_stub
        service.notify("Test Email Message")

        self.assertIn(
            "Test Email Message",
            email_stub.messages_sent,
        )
```

We need another function for the notification with SMS functionality, `test_notify_with_sms`. Similarly to the previous case, we create an instance of `SMSSenderStub`. Then, we need to inject that stub into the notification service. But, for that, in the scope of the test, we define a custom notification service class, and decorate it with `@inject_sender(SMSSenderStub)`, as follows:

```
@inject_sender(SMSSenderStub)
class CustomNotificationService:
    sender: NotificationSender = None

    def notify(self, message):
        self.sender.send(message)
```

Based on that, we inject the SMS sender stub into the custom service, send a notification message, and then verify that the message was sent by the SMS stub. The complete code for the second unit test is as follows:

```
    def test_notify_with_sms(self):
        sms_stub = SMSSenderStub()

        @inject_sender(SMSSenderStub)
        class CustomNotificationService:
            sender: NotificationSender = None

            def notify(self, message):
                self.sender.send(message)

        service = CustomNotificationService()
        service.sender = sms_stub
        service.notify("Test SMS Message")

        self.assertIn(
            "Test SMS Message", sms_stub.messages_sent
        )
```

Finally, we should not forget to add the lines needed for executing unit tests when the file is interpreted by Python:

```
if __name__ == "__main__":
    unittest.main()
```

Executing the unit test code (in the `ch10/dependency_injection/test_di_with_decorator.py` file), using the `python ch10/dependency_injection/test_di_with_decorator.py` command, gives the following output:

```
..
-------------------------------------------------------------
Ran 2 tests in 0.000s

OK
```

This is what was expected.

So, this example showed how using a decorator to manage dependencies allows for easy changes without modifying the class internals, which not only keeps the application flexible but also encapsulates the dependency management outside of the core business logic of your application. In addition, we saw how DI can be tested with unit tests using the stubs technique, ensuring the application's components work as expected in isolation.

Summary

In this chapter, we've explored two pivotal patterns essential for writing clean code and enhancing our testing strategies: the Mock Object pattern and the Dependency Injection pattern.

The Mock Object pattern is crucial for ensuring test isolation, which helps avoid unwanted side effects. It also facilitates behavior verification and simplifies test setup. We discussed how mocking, particularly through the `unittest.mock` module, allows us to simulate components within a unit test, demonstrating this with a practical example.

The Dependency Injection pattern, on the other hand, offers a robust framework for managing dependencies in a way that enhances flexibility, testability, and maintainability. It's applicable not only in testing scenarios but also in general software design. We illustrated this pattern with an initial example that integrates mocking for either unit or integration tests. Subsequently, we explored a more advanced implementation using a decorator to streamline dependency management across both the application and its tests.

As we conclude this chapter and prepare to enter the final one, we'll shift our focus slightly to discuss Python anti-patterns, identifying common pitfalls, and learning how to avoid them.

11

Python Anti-Patterns

In this final chapter, we will explore Python anti-patterns. These are common programming practices that, while not necessarily wrong, often lead to less efficient, less readable, and less maintainable code. By understanding these pitfalls, you can write cleaner, more efficient code for your Python applications.

In this chapter, we're going to cover the following main topics:

- Code style violations
- Correctness anti-patterns
- Maintainability anti-patterns
- Performance anti-patterns

Technical requirements

See the requirements presented in *Chapter 1.*

Code style violations

The Python style guide, also known as **Python Enhancement Proposal no 8** (**PEP 8**), provides recommendations for readability and consistency in your code, making it easier for developers to collaborate and maintain projects over time. You can find the style guide details on its official page here: https://peps.python.org/pep-0008. In this section, we are going to present some of the recommendations of the style guide so that you can avoid them when writing your application's or library's code.

Tools for fixing coding style violations

Note that we have formatting tools such as *Black* (`https://black.readthedocs.io/en/stable/`), *isort* (`https://pycqa.github.io/isort/`), and/or *Ruff* (`https://docs.astral.sh/ruff/`) that can help you fix code that does not follow the style guide recommendations. We are not going to spend time on how to use these tools here since you can find all the needed documentation on their official documentation pages and start using them in a matter of minutes.

Now, let's explore our selected code style recommendations.

Indentation

You should use four spaces per indentation level, and you should avoid mixing tabs and spaces.

Maximum line length and blank lines

The style guide recommends limiting all lines of code to a maximum of 79 characters, for better readability.

Also, there are rules related to blank lines. First, you should surround top-level function and class definitions with two blank lines. Second, method definitions inside a class should be surrounded by a single blank line.

For example, the formatting with the following code snippet is incorrect:

```
class MyClass:
    def method1(self):
        pass
    def method2(self):
        pass
def top_level_function():
    pass
```

The right formatting is as follows:

```
class MyClass:

    def method1(self):
        pass

    def method2(self):
        pass


def top_level_function():
    pass
```

Imports

The way you write, organize, and order your import lines is also important. According to the style guide, imports should be on separate lines and grouped into three categories in this order: standard library imports, related third-party imports, and local-specific imports within the application's or library's code base. Also, each group should be separated by a blank line.

For example, the following is not compliant with the style guide:

```
import os, sys
import numpy as np
from mymodule import myfunction
```

The best practice for the same imports is the following:

```
import os
import sys

import numpy as np

from mymodule import myfunction
```

Naming conventions

You should use descriptive names for variables, functions, classes, and modules. The following are specific naming conventions for different types of cases:

- **Name for function and variable (including class attributes and methods)**: Use `lower_case_with_underscores`
- **Name for class**: Use `CapWords`
- **Name for constant**: Use `ALL_CAPS_WITH_UNDERSCORES`

For example, the following is not good practice:

```
def calculateSum(a, b):
    return a + b

class my_class:
    pass

maxValue = 100
```

The best practice is the following:

```
def calculate_sum(a, b):
    return a + b
```

```
class MyClass:
    pass

MAX_VALUE = 100
```

Comments

Comments should be complete sentences, with the first word capitalized, and should be clear and concise. We have specific recommendations for two cases of comments—block comments and inline comments:

- Block comments generally apply to some (or all) code that follows them and are indented to the same level as that code. Each line of a block comment starts with # and a single space.

- Inline comments should be used sparingly. An inline comment is placed on the same line as a statement, separated by at least two spaces from the statement.

For example, in the following, we have a bad comment style:

```
#This is a poorly formatted block comment.
def foo():  #This is a poorly formatted inline comment.
    pass
```

Here is the equivalent code, with the style fixed:

```
# This is a block comment.
# It spans multiple lines.
def foo():
    pass  # This is an inline comment.
```

Whitespace in expressions and statements

You should avoid extraneous whitespace in the following situations:

- Immediately inside parentheses, brackets, or braces

- Immediately before a comma, semicolon, or colon

- More than one space around an assignment operator to align it with another

This ends our review of the most common code style violations to pay attention to. As previously said, there are tools to detect and fix such violations in your code in a productive way, and they are generally included in the developer workflow (for example, via git commit hooks and/or in the project's CI/CD processes).

Correctness anti-patterns

These anti-patterns can lead to bugs or unintended behavior if not addressed. We are going to discuss the most common of these anti-patterns and alternative, recommended ways and approaches. We are going to focus on the following anti-patterns:

- Using the type() function for comparing types

- Mutable default argument

- Accessing a protected member from outside a class

Note that using IDEs such as *Visual Studio Code* or *PyCharm* or command-line tools such as *Flake8* will help you spot such bad practices in your code, but it is important to know the recommendations and the reason behind each one.

Using the type() function for comparing types

Sometimes, we need to identify the type of a value through comparison, for our algorithm. The common technique one may think of for that is to use the type() function. But using type() to compare object types does not account for subclassing and is not as flexible as the alternative which is based on using the isinstance() function.

Imagine we have two classes, CustomListA and CustomListB, that are subclasses of the UserList class, which is the recommended class one should inherit from when defining a class for a custom list, as follows:

```python
from collections import UserList


class CustomListA(UserList):
    pass


class CustomListB(UserList):
    pass
```

If we wanted to check if an object is of one of the custom list types, using the first approach, we would test the type(obj) in (CustomListA, CustomListB) condition.

Alternatively, we would simply test isinstance(obj, UserList), and that would be enough since CustomListA and CustomListB are subclasses of UserList.

As a demonstration, we write a `compare()` function that uses the first approach, as follows:

```
def compare(obj):
    if type(obj) in (CustomListA, CustomListB):
        print("It's a custom list!")
    else:
        print("It's a something else!")
```

Then, we write a `better_compare()` function to do the equivalent using the alternative approach, as follows:

```
def better_compare(obj):
    if isinstance(obj, UserList):
        print("It's a custom list!")
    else:
        print("It's a something else!")
```

The following lines of code can help test both functions:

```
obj1 = CustomListA([1, 2, 3])
obj2 = CustomListB(["a", "b", "c"])

compare(obj1)
compare(obj2)
better_compare(obj1)
better_compare(obj2)
```

The complete demonstration code is in the `ch11/compare_types.py` file. Running the `python ch11/compare_types.py` command should give the following output:

```
It's a custom list!
It's a custom list!
It's a custom list!
It's a custom list!
```

This shows that both functions can produce the expected result. But the function using the recommended technique, `isinstance()`, is simpler to write and more flexible since it takes subclasses into account.

Mutable default argument

When you define a function with a parameter that expects a mutable value, such as a list or a dictionary, you may be tempted to provide a default argument (`[]` or `{}` respectively). But such a function retains changes between calls, which will lead to unexpected behaviors.

The recommended practice is to use a default value of None and set it to a mutable data structure within the function if needed.

Let's create a function called manipulate() whose mylist parameter has a default value of []. The function appends the "test" string to the mylist list and then returns it, as follows:

```
def manipulate(mylist=[]):
    mylist.append("test")
    return mylist
```

In another function called better_manipulate() whose mylist parameter has a default value of None, we start by setting mylist to [] if it is None, then we append the "test" string to mylist before returning it, as follows:

```
def better_manipulate(mylist=None):
    if not mylist:
        mylist = []

    mylist.append("test")
    return mylist
```

The following lines help us test each function by calling it several times with the default argument:

```
if __name__ == "__main__":
    print("function manipulate()")
    print(manipulate())
    print(manipulate())
    print(manipulate())

    print("function better_manipulate()")
    print(better_manipulate())
    print(better_manipulate())
```

Running the python ch11/mutable_default_argument.py command should give the following output:

```
function manipulate()
['test']
['test', 'test']
['test', 'test', 'test']
function better_manipulate()
['test']
['test']
```

As you can see, with the non-recommended way of doing this, we end up with the `"test"` string several times in the list returned; the string is accumulating because each subsequent time the function has been called, the `mylist` argument kept its previous value instead of being reset to the empty list. But, with the recommended solution, we see with the result that we get the expected behavior.

Accessing a protected member from outside a class

Accessing a protected member (an attribute prefixed with _) of a class from outside that class usually calls for trouble since the creator of that class did not intend this member to be exposed. Someone maintaining the code could change or rename that attribute later down the road, and parts of the code accessing it could result in unexpected behavior.

If you have code that accesses a protected member that way, the recommended practice is to refactor that code so that it is part of the public interface of the class.

To demonstrate this, let's define a Book class with two protected attributes, _title and _author, as follows:

```
class Book:
    def __init__(self, title, author):
        self._title = title
        self._author = author
```

Now, let's create another class, BetterBook, with the same attributes and a `presentation_line()` method that accesses the _title and _author attributes and returns a concatenated string based on them. The class definition is as follows:

```
class BetterBook:
    def __init__(self, title, author):
        self._title = title
        self._author = author

    def presentation_line(self):
        return f"{self._title} by {self._author}"
```

Finally, in the code for testing both classes, we get and print the presentation line for an instance of each class, accessing the protected members for the first one (instance of Book) and calling the `presentation_line()` method for the second one (instance of BetterBook), as follows:

```
if __name__ == "__main__":
    b1 = Book(
        "Mastering Object-Oriented Python",
        "Steven F. Lott",
    )
```

```
print(
    "Bad practice: Direct access of protected members"
)
print(f"{b1._title} by {b1._author}")

b2 = BetterBook(
    "Python Algorithms",
    "Magnus Lie Hetland",
)
print(
    "Recommended: Access via the public interface"
)
print(b2.presentation_line())
```

The complete code is in the ch11/ protected_member_of_class.py file. Running the python ch11/ protected_member_of_class.py command gives the following output:

```
Bad practice: Direct access of protected members
Mastering Object-Oriented Python by Steven F. Lott
Recommended: Access via the public interface
Python Algorithms by Magnus Lie Hetland
```

This shows that we get the same result, without any error, in both cases, but using the presentation_ line() method, as done in the case of the second class, is the best practice. The _title and _author attributes are protected, so it is not recommended to call them directly. The developer could change those attributes in the future. That is why they must be encapsulated in a public method.

Also, it is good practice to provide an attribute that encapsulates each protected member of the class using the @property decorator, as we have seen in the *Techniques for achieving encapsulation* section of *Chapter 1, Foundational Design Principles*.

Maintainability anti-patterns

These anti-patterns make your code difficult to understand or maintain over time. We are going to discuss several anti-patterns that should be avoided for better quality in your Python application or library's code base. We will focus on the following points:

- Using a wildcard import
- **Look Before You Leap (LBYL)** versus **Easier to Ask for Forgiveness than Permission (EAFP)**
- Overusing inheritance and tight coupling
- Using global variables for sharing data between functions

As mentioned for the previous category of anti-patterns, using tools such as Flake8 as part of your developer workflow can be handy to help find some of those potential issues when they are already present in your code.

Using a wildcard import

This way of importing (`from mymodule import *`) can clutter the namespace and make it difficult to determine where an imported variable or function came from. Also, the code may end up with bugs because of name collision.

The best practice is to use specific imports or import the module itself to maintain clarity.

LBYL versus EAFP

LBYL often leads to more cluttered code, while EAFP makes use of Python's handling of exceptions and tends to be cleaner.

For example, we may want to check if a file exists, before opening it, with code such as the following:

```
if os.path.exists(filename):
    with open(filename) as f:
        print(f.text)
```

This is LBYL, and when new to Python, you would think that it is the right way to treat such situations. But in Python, it is recommended to favor EAFP, where appropriate, for cleaner, more Pythonic code. So, the recommended way for the expected result would give the following code:

```
try:
    with open(filename) as f:
        print(f.text)
except FileNotFoundError:
    print("No file there")
```

As a demonstration, let's write a `test_open_file()` function that uses the LBYL approach, as follows:

```
def test_open_file(filename):
    if os.path.exists(filename):
        with open(filename) as f:
            print(f.text)
    else:
        print("No file there")
```

Then, we add a function that uses the recommended approach:

```
def better_test_open_file(filename):
    try:
        with open(filename) as f:
            print(f.text)
    except FileNotFoundError:
        print("No file there")
```

We can then test these functions with the following code:

```
filename = "no_file.txt"
test_open_file(filename)
better_test_open_file(filename)
```

You can check the complete code of the example in the ch11/lbyl_vs_eafp.py file, and running it should give the following output:

```
No file there
No file there
```

This output shows us that both approaches give the same outcome, but the try/except way makes our code cleaner.

Overusing inheritance and tight coupling

Inheritance is a powerful feature of OOP, but overusing it – for example, creating a new class for every slight variation of behavior – can lead to tight coupling between classes. This increases complexity and makes the code less flexible and harder to maintain.

It is not recommended to create a deep inheritance hierarchy such as the following (as a simplified example):

```
class GrandParent:
    pass


class Parent(GrandParent):
    pass


class Child(Parent):
    Pass
```

The best practice is to create smaller, more focused classes and combine them to achieve the desired behavior, as with the following:

```
class Parent:
    pass
```

```
class Child:
    def __init__(self, parent):
        self.parent = parent
```

As you may remember, this is the composition approach, which we discussed in the *Following the Favor Composition over Inheritance principle* section of *Chapter 1, Foundational Design Principles*.

Using global variables for sharing data between functions

Global variables are variables that are accessible throughout the entire program, making them tempting to use for sharing data between functions—for example, configuration settings that are used across multiple modules or shared resources such as database connections.

However, they can lead to bugs where different parts of the application unexpectedly modify global state. Also, they make it harder to scale applications as they can lead to issues in multithreaded environments where multiple threads might attempt to modify the global variable simultaneously.

Here is an example of the non-recommended practice:

```
# Global variable
counter = 0

def increment():
    global counter
    counter += 1

def reset():
    global counter
    counter = 0
```

Instead of using a global variable, you should pass the needed data as arguments to functions or encapsulate state within a class, which improves the modularity and testability of the code. So, the best-practice equivalent for the counterexample would be defining a Counter class holding a counter attribute, as follows:

```
class Counter:
    def __init__(self):
        self.counter = 0

    def increment(self):
        self.counter += 1

    def reset(self):
        self.counter = 0
```

Next, we add code for testing the Counter class as follows:

```
if __name__ == "__main__":
    c = Counter()
    print(f"Counter value: {c.counter}")
    c.increment()
    print(f"Counter value: {c.counter}")
    c.reset()
```

You can check the complete code of the example in the ch11/instead_of_global_variable. py file, and running it should give the following output:

```
Counter value: 0
Counter value: 1
```

This shows how using a class instead of global variables is effective and can be scalable, thus the recommended practice.

Performance anti-patterns

These anti-patterns lead to inefficiencies that can degrade performance, especially noticeable in large-scale applications or data-intensive tasks. We will focus on the following such anti-patterns:

- Not using .join() to concatenate strings in a loop

- Using global variables for caching

Let's start.

Not using .join() to concatenate strings in a loop

Concatenating strings with + or += in a loop creates a new string object each time, which is inefficient. The best solution is to use the .join() method on strings, which is designed for efficiency when concatenating strings from a sequence or iterable.

Let's create a concatenate() function where we use += for concatenating items from a list of strings, as follows:

```
def concatenate(string_list):
    result = ""
    for item in string_list:
        result += item
    return result
```

Then, let's create a `better_concatenate()` function for the same result, but using the `str.join()` method, as follows:

```
def better_concatenate(string_list):
    result = "".join(string_list)
    return result
```

We can then test both functions using the following:

```
if __name__ == "__main__":
    string_list = ["Abc", "Def", "Ghi"]
    print(concatenate(string_list))
    print(better_concatenate(string_list))
```

Running the code (in the `ch11/concatenate_strings_in_loop.py` file) gives the following output:

```
AbcDefGhi
AbcDefGhi
```

This confirms that both techniques produce the same result, though using `.join()` is the recommended practice for performance reasons.

Using global variables for caching

Using global variables for caching can seem like a quick and easy solution but often leads to poor maintainability, potential data consistency issues, and difficulties in managing the cache life cycle effectively. A more robust approach involves using specialized caching libraries designed to handle these aspects more efficiently.

In this example (in the `ch11/caching/using_global_var.py` file), a global dictionary is used to cache results from a function that simulates a time-consuming operation (for example, a database query) done in the `perform_expensive_operation()` function. The complete code for this demonstration is as follows:

```
import time
import random

# Global variable as cache
_cache = {}

def get_data(query):
    if query in _cache:
        return _cache[query]
    else:
```

```
        result = perform_expensive_operation(query)
        _cache[query] = result
        return result


def perform_expensive_operation(user_id):

    time.sleep(random.uniform(0.5, 2.0))

    user_data = {
        1: {"name": "Alice", "email": "alice@example.com"},
        2: {"name": "Bob", "email": "bob@example.com"},
        3: {"name": "Charlie", "email": "charlie@example.com"},
    }

    result = user_data.get(user_id, {"error": "User not found"})

    return result


if __name__ == "__main__":
    print(get_data(1))
    print(get_data(1))
```

Testing the code by running the `python ch11/caching/using_global_var.py` command gives the following output:

```
{'name': 'Alice', 'email': 'alice@example.com'}
{'name': 'Alice', 'email': 'alice@example.com'}
```

This works as expected, but there is a better approach. We can use a specialized caching library or Python's built-in `functools.lru_cache()` function. The `lru_cache` decorator provides a **least recently used (LRU)** cache, automatically managing the size and lifetime of cache entries. Also, it is thread-safe, which helps prevent issues that can arise in a multithreaded environment when multiple threads access or modify the cache simultaneously. Finally, libraries or tools such as `lru_cache` are optimized for performance, using efficient data structures and algorithms to manage the cache.

Here's how you can implement the functionality of caching results from a time-consuming function using `functools.lru_cache`. The complete code (in the `ch11/caching/using_lru_cache.py` file) is as follows:

```
import random
import time
from functools import lru_cache
```

```
@lru_cache(maxsize=100)
def get_data(user_id):
    return perform_expensive_operation(user_id)


def perform_expensive_operation(user_id):
    time.sleep(random.uniform(0.5, 2.0))

    user_data = {
        1: {"name": "Alice", "email": "alice@example.com"},
        2: {"name": "Bob", "email": "bob@example.com"},
        3: {"name": "Charlie", "email": "charlie@example.com"},
    }

    result = user_data.get(user_id, {"error": "User not found"})

    return result


if __name__ == "__main__":
    print(get_data(1))
    print(get_data(1))
    print(get_data(2))
    print(get_data(99))
```

To test this code, run the `python ch11/caching/using_lru_cache.py` command. You should get the following output:

```
{'name': 'Alice', 'email': 'alice@example.com'}
{'name': 'Alice', 'email': 'alice@example.com'}
{'name': 'Bob', 'email': 'bob@example.com'}
{'error': 'User not found'}
```

As we can see, this approach not only enhances the robustness of the caching mechanism but also improves code readability and maintainability.

Summary

Understanding and avoiding common Python anti-patterns will help you write cleaner, more efficient, and maintainable code.

First, we presented common Python code style violations. Then we discussed several anti-patterns that are related to correctness and can lead to bugs. Next, we covered practices that, beyond the code style itself, are not good for code readability and maintainability. Finally, we saw a couple of anti-patterns that one should avoid for writing code that has good performance.

Always remember – the best code is not just about making it work but also about making it work well. Even more, ideally, it should be easy to maintain.

We finally reached the end of this book. It was quite a journey. We started with the main design principles, then moved on to cover the most popular design patterns in the way they can be applied to Python, and finally touched upon Python anti-patterns. That's a lot! The ideas and examples we discussed help us to think about different implementation options or techniques to choose from whenever we have a use case. Whatever the solution you choose, keep in mind that Python favors simplicity, try to use patterns and techniques that are considered Pythonic, and avoid Python's anti-patterns.

Index

packtpub.com

Subscribe to our online digital library for full access to over 7,000 books and videos, as well as industry leading tools to help you plan your personal development and advance your career. For more information, please visit our website.

Why subscribe?

- Spend less time learning and more time coding with practical eBooks and Videos from over 4,000 industry professionals

- Improve your learning with Skill Plans built especially for you

- Get a free eBook or video every month

- Fully searchable for easy access to vital information

- Copy and paste, print, and bookmark content

Did you know that Packt offers eBook versions of every book published, with PDF and ePub files available? You can upgrade to the eBook version at packtpub.com and as a print book customer, you are entitled to a discount on the eBook copy. Get in touch with us at customercare@packtpub.com for more details.

At www.packtpub.com, you can also read a collection of free technical articles, sign up for a range of free newsletters, and receive exclusive discounts and offers on Packt books and eBooks.

Other Books You May Enjoy

If you enjoyed this book, you may be interested in these other books by Packt:

Advanced Python Programming

Quan Nguyen

ISBN: 978-1-80181-401-0

- Write efficient numerical code with NumPy, pandas, and Xarray
- Use Cython and Numba to achieve native performance
- Find bottlenecks in your Python code using profilers
- Optimize your machine learning models with JAX
- Implement multithreaded, multiprocessing, and asynchronous programs
- Solve common problems in concurrent programming, such as deadlocks
- Tackle architecture challenges with design patterns

Metaprogramming with Python

Sulekha AloorRavi

ISBN: 978-1-83855-465-1

- Understand the programming paradigm of metaprogramming and its need
- Revisit the fundamentals of object-oriented programming
- Define decorators and work with metaclasses
- Employ introspection and reflection on your code
- Apply generics, typing, and templates to enhance your code
- Get to grips with the structure of your code through abstract syntax trees and the behavior through method resolution order
- Create dynamic objects and generate dynamic code
- Understand various design patterns and best practices

Packt is searching for authors like you

If you're interested in becoming an author for Packt, please visit `authors.packtpub.com` and apply today. We have worked with thousands of developers and tech professionals, just like you, to help them share their insight with the global tech community. You can make a general application, apply for a specific hot topic that we are recruiting an author for, or submit your own idea.

Share your thoughts

Now you've finished *Mastering Python Design Patterns*, we'd love to hear your thoughts! Scan the QR code below to go straight to the Amazon review page for this book and share your feedback or leave a review on the site that you purchased it from.

https://packt.link/r/1837639612

Your review is important to us and the tech community and will help us make sure we're delivering excellent quality content.

Download a free PDF copy of this book

Thanks for purchasing this book!

Do you like to read on the go but are unable to carry your print books everywhere?

Is your eBook purchase not compatible with the device of your choice?

Don't worry, now with every Packt book you get a DRM-free PDF version of that book at no cost.

Read anywhere, any place, on any device. Search, copy, and paste code from your favorite technical books directly into your application.

The perks don't stop there, you can get exclusive access to discounts, newsletters, and great free content in your inbox daily

Follow these simple steps to get the benefits:

1. Scan the QR code or visit the link below

https://packt.link/free-ebook/9781837639618

2. Submit your proof of purchase
3. That's it! We'll send your free PDF and other benefits to your email directly

www.ingramcontent.com/pod-product-compliance
Lightning Source LLC
Chambersburg PA
CBHW080629060326
40690CB00021B/4863